Encyclopedia of Alternative and Renewable Energy: Thin Film Solar Cells

Volume 24

Encyclopedia of Alternative and Renewable Energy: Thin Film Solar Cells

Volume 24

Edited by **Terence Maran and**
David McCartney

New York

Published by Callisto Reference,
106 Park Avenue, Suite 200,
New York, NY 10016, USA
www.callistoreference.com

Encyclopedia of Alternative and Renewable Energy: Thin Film Solar Cells
Volume 24
Edited by Terence Maran and David McCartney

International Standard Book Number: 978-1-63239-198-8 (Hardback)

Printed in the United States of America.

Contents

Preface

This book discusses the benefits and challenges of utilizing thin film solar cells as an alternative energy source. The field of photovoltaics has seen a large-scale manufacturing of the second genesis of thin film solar modules and has succeeded in constructing powerful solar plants in many countries across the globe. Thin film techniques using direct-gap semiconductors such as CIGS and CdTe pose minimum manufacturing costs and are now increasing in popularity amongst industries. This has led to an increase in the manufacturability of thin film solar modules as compared to wafer or ribbon Si modules. Thin films like CIGS and CdTe will soon take over wafer-based silicon solar cells as the superior photovoltaic technology. This book elucidates the scientific and technological difficulties of increasing the photoelectric efficiency of thin film solar cells. It covers various aspects of thin film solar cells varying from photovoltaics as mainstream power engineering to low cost solar cell based on cuprous oxides to application of electron beam treatment. This book will be beneficial for readers interested in this subject.

This book is a comprehensive compilation of works of different researchers from varied parts of the world. It includes valuable experiences of the researchers with the sole objective of providing the readers (learners) with a proper knowledge of the concerned field. This book will be beneficial in evoking inspiration and enhancing the knowledge of the interested readers.

In the end, I would like to extend my heartiest thanks to the authors who worked with great determination on their chapters. I also appreciate the publisher's support in the course of the book. I would also like to deeply acknowledge my family who stood by me as a source of inspiration during the project.

Editor

Thin-Film Photovoltaics
as a Mainstream of Solar Power Engineering

Leonid A. Kosyachenko
Chernivtsi National University
Ukraine

1. Introduction

Provision of energy is one of the most pressing problems facing humanity in the 21st century. Without energy, it is impossible to overcome the critical issues of our time. Industrial world suggests continuous growth in energy consumption in the future.

According to the U.S. Department of Energy, the world's generating capacity is now close to 18 TW. The main source of energy even in highly developed countries is fossil fuel, i.e. coal, oil and natural gas. However, resources of fossil fuel are limited, and its production and consumption irreversibly affect the environmental conditions with the threat of catastrophic climate change on Earth. Other energy sources, particularly *nuclear* energy, are also used that would fully meet *in principle* the energy needs of mankind. Capacity of existing nuclear reactors (nearly 450 in the world) is ~ 370 GW. However, increasing their capacity up to ~ 18 TW or about 50 times (!), is quite problematic (to provide humanity with *electric* energy, the capacity of nuclear power should be increased about 10 times). Resources of hydroelectric, geothermal, wind energy, energy from biofuels are also limited. At the same time, the power of solar radiation of the Earth's surface exceeds the world's generating capacity by more than 1000 times. It remains only to master this accessible, inexhaustible, gratuitous and nonhazardous source of energy in an environmentally friendly way.

Solar energy can be converted into heat and electricity. Different ways of converting sunlight into electricity have found practical application. The power plants, in which water is heated by sunlight concentrating devices resulting in a high-temperature steam and operation of an electric generator, are widespread. However, solar cells are much more attractive due to the *direct* conversion of solar radiation into electricity. This is the so-called *photovoltaics.* Under the conditions of the growing problems of global warming, photovoltaics is the most likely candidate to replace fossil fuels and nuclear reactors.

2. Silicon solar cells

Over the decades, solar modules (panels) based on single-crystalline (mono-crystalline, c-Si), polycrystalline (multi-crystalline, mc-Si), ribbon (ribbon-Si) and amorphous (a-Si) silicon are dominant in photovoltaics (Fig. 1).

In recent years, photovoltaics demonstrates high growth rates in the entire energy sector. According to the European Photovoltaic Industry Association , despite the global financial and economic crisis, the capacity of installed solar modules in the world grew by 16.6 GW in

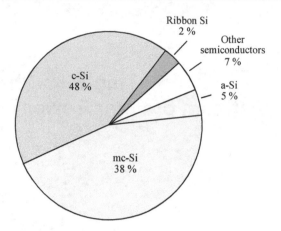

Fig. 1. Distribution of capacity of photovoltaic energy in the world.

2010, and their total capacity reached ~ 40 GW, that is almost 8 times more than in 2005. The growth rate of the photovoltaic energy for the next 4-5 years is expected to be quite high. In 2014, capacity of installed modules will be about 14 GW and 30 GW according to the pessimistic and optimistic forecast, respectively. Nevertheless, despite the relatively high annual growth, the contribution of semiconductor solar cells in the global energy system is small (less than 0.3%!), and the prospects for *desired rapid* development of photovoltaics are not reassuring (Fig. 2). The contribution of Si photovoltaic solar power plants in generation capacity in the world will reach ~ 1% only in the years 2018-2020, and may exceed 10% in the years 2045-2050 (EPIA, 2011; EUR 24344 PV, 2010; Jager-Waldau, 2010). Analysts do not accept the development of the PV scenario shown by the dashed line in Fig. 2. Thus, solving the energy problems by developing Si photovoltaics seems too lengthy.[1]

The reason for the slow power growth of traditional silicon solar modules lies in a large consumption of materials and energy, high labor intensiveness and, as a consequence, a low productivity and high cost of modules with acceptable photovoltaic conversion efficiency for mass production (16-17 and 13-15% in the case of single-crystalline and polycrystalline material, respectively) (Szlufcik et al., 2003; Ferrazza, 2003).[2] The problem is fundamental and lies in the fact that silicon is an *indirect* semiconductor and therefore the *total* absorption needs its significant thickness (up to 0.5 mm and more). As a result, to collect the charge photogenerated in a thick absorbing layer, considerable diffusion length of minority carriers (long lifetime and high mobility) and, therefore, high quality material with high *carrier diffusion length of hundreds of micrometers* are required.

Estimating the required thickness of the semiconductor in solar cells, one is often guided by an effective penetration depth of radiation into the material α^{-1}, where α is the absorption coefficient in the region of electronic interband transitions. However, the value of α varies

---

[1] In the European Union, these rates are much higher. Now the cumulative power of solar modules is 1.2% and, by 2015 and 2020, will rise to 4-5% and 6-12% of the EU's electricity demand, respectively.
[2] A lot of effort has been undertaken to increase the efficiency of silicon solar cells above 20-24% but improvements are reached only with the help of cost-intensive processes, which usually cannot be implemented into industrial products (Koch et al., 2003).

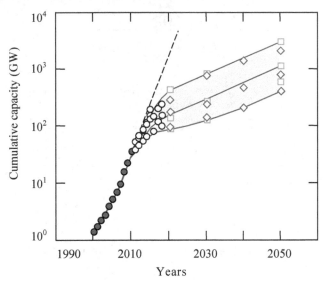

Fig. 2. Evolution of world cumulative installed PV capacity until 2050: ● historical data, ○, □, ◇ forecasts (EPIA, 2011; EUR 24344 PV, 2010; Hegedus & Luque, 2011).

rather widely, especially in the indirect semiconductor, and solar radiation is distributed over the spectrum in a complicated manner (Fig. 3(a)). Therefore, the absorptive capacity (absorptivity) of the material, used in solar cell, can be described by a certain *integral* characteristic, which takes into account the absorption spectrum of the material and the spectral distribution of solar radiation. For a structure with flat surfaces, the integral absorption ability of the radiation, which has penetrated into the material (certain part of radiation is reflected from the front surface), can be represented as

$$A(d) = \frac{\sum_i \frac{\Phi_i + \Phi_{i-1}}{2}\left[1 - \exp\left(-\frac{\alpha_i + \alpha_{i-1}}{2}d\right)\right]\Delta\lambda_i}{\sum_i \frac{\Phi_i + \Phi_{i-1}}{2}\Delta\lambda_i}, \tag{1}$$

where Φ_i is the spectral power density of solar radiation at the wavelength λ_i under standard solar irradiation AM1.5 shown in Fig. 3(a), $\Delta\lambda_i$ is the spacing between neighboring wavelengths in Table 9845-1 of the International Organization for Standardization (Standard ISO, 1992), and α_i is the absorption coefficient at wavelength λ_i. The summation in Eq. (1) is made from $\lambda \approx 300$ nm to $\lambda = \lambda_g = hc/E_g$, since at wavelengths λ shorter than 300 nm, terrestrial radiation of the Sun is virtually absent, and when $\lambda > \lambda_g$, radiation is not absorbed in the material with the generation of electron-hole pairs.

Fig. 3(b) shows the dependences of absorptivity of solar radiation A of single-crystalline silicon on the thickness of the absorber layer d calculated by Eq. (1).

As seen in Fig. 3(b), in crystalline silicon, the total absorption of solar radiation in the fundamental absorption region ($h\nu \geq E_g$) occurs when d is close to 1 cm (!), and 95% of the radiation is absorbed at a thickness of about 300 μm. More absorption can be achieved by using the reflection of light from the rear surface of the solar cell, which is usually completely

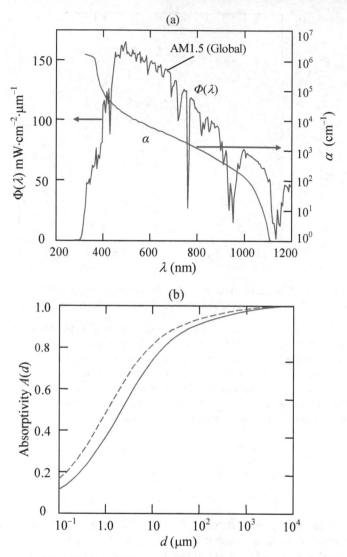

Fig. 3. (a) – Power spectral density of the total solar radiation Φ under AM1.5 conditions and the absorption curve $\alpha(\lambda)$ for crystalline silicon. (b) – Dependence of absorptivity of solar radiation in the $h\nu \geq E_g$ spectral range on the absorber layer thickness d for crystalline silicon. The dashed line shows the absorptivity of a silicon wafer taking into account 100% reflection from its rear surface.

covered with metal. If in the ideal case, the reflectivity of light from the rear surface is unity, the absorption of the plate will be such as if its thickness is twice as much. In this case, 95% of the radiation is absorbed by the plate of 150 μm thickness.

Of course, it has to be rejected to use silicon wafers of thickness a few millimeters, so that the absorption of solar radiation was complete. Many companies producing silicon modules agreed on a compromise thickness of 150-250 μm, when about 93% of solar radiation with photon energy $hv \geq E_g$ is absorbed or about 94% when the rear surface of the solar cell is mirror (Szlufcik et al., 2003; Ferrazza, 2003).[3] Deficiency of absorption in the material offsets by the creation of a special profile on surfaces (texturing) and by other ways. Needless to say, an anti-reflective coating is applied to reduce significantly the reflection from the front surface because over 30% of the radiation is reflected from a flat silicon surface.

Production of solar modules based on silicon wafers involves a lot of stages (Hegedus & Luque, 2011). The so-called metallurgical grade (MG) silicon is obtained from quarzite (SiO_2) with charcoal in a high-temperature arc furnace. Then MG silicon is highly purified commonly by a method developed by the Siemens Company consisting of the fractional distillation of chlorosilanes. Finally, chlorosilanes are reduced with hydrogen at high temperatures to produce the so-called semiconductor grade (SG) silicon. By recrystallizing such polycrystalline silicon, single-crystalline Si ingots are often grown by the Czochralski (Cz) or the floating-zone (FZ) techniques adapted from the microelectronics industry. This is followed by cutting (slicing) the ingot into wafers, of course, with considerable waste. It should be noted that the cost of silicon purification, production of ingots, slicing them into wafers constitute up to 40-55% of the cost of solar module.

Manufacture of conventional silicon solar cell also includes a number of other operations. Among them, (i) chemical etching wafers to provide removal of the layer damaged during slicing and polishing; (ii) high-temperature diffusion to create a p-n junction; (iii) anisotropic etching to build a surface structure with random pyramids that couples the incoming light more effectively into the solar cell; (iv) complicated procedure of applying full area and grid-like ohmic contacts to p- and n-type regions provided a minimum of electrical and recombination losses (contacts in silicon solar cells are often made by screen-printing metal paste, which is then annealed at several hundred degrees Celsius to form metal electrodes), etc. (Mauk, et al., 2003). Once the cells are manufactured they are assembled into modules either in the cell factories or in module assembly factories that purchase cells from variety cell factories (Hegedus & Luque, 2011). All that complicates the manufacturing technology and, hence, reduces the productivity and increases the cost of solar modules.

Summing up, one should again emphasize that single-crystalline Si modules are among the most efficient but at the same time the most expensive since they require the highest purity silicon and involve a lot of stages of complicated processes in their manufacture.

For decades, an intensive search for cheaper production technology of silicon solar cells is underway. Back in the 1980's, a technology of material solidification processes for production of large silicon ingots (blocks with weights of 250 to 300 kg) of **polycrystalline (multicrystalline) silicon** (mc-Si) has been developed (Koch et al., 2003). In addition to lower cost manufacturing process, an undoubted advantage of mc-Si is the rational use of the material in the manufacture of solar cells due to the rectangular shape of the ingot. In the case of a single-crystalline ingot of cylindrical form, the so-called "pseudo-square" wafers with rounded corners are used, i.e. c-Si modules have some gaps at the four corners of the cells).

Polycrystalline silicon is characterized by defects caused by the presence of random grains of crystalline Si, a significant concentration of dislocations and other crystal defects (impurities). These defects reduce the carrier lifetime and mobility, enhance recombination of carriers and ultimately decrease the solar cells efficiency. Thus, polysilicon-based cells

---

[3] Further thinning of silicon is also constrained by the criteria of mechanical strength of a wafer as well as the handling and processing techniques (silicon is brittle).

are less expensive to produce single-crystalline silicon cells but are less efficient. As a result, cost per unit of generated electric power ("specific" or "relative" cost) for c-Si and mc-Si modules is practically equal (though the performance gap has begun to close in recent years). The polysilicon-based cells are the most common solar modules on the market being less expensive than single-crystalline silicon.

A number of methods for growing the so-called **ribbon-Si**, i.e. a polycrystalline silicon in the form of thin sheets, is also proposed. The advantages of ribbon silicon are obvious, as it excludes slicing the ingot into thin wafers, allowing material consumption to reduce roughly *halved*. However, the efficiency of ribbon-Si solar cells is not as high as of mc-Si cells because the need of high quality material with the thickness of the absorber layer of 150-250 μm and, hence, with high carrier diffusion lengths of hundreds of micrometers remains (Hegedus & Luque, 2011). Nevertheless, having a lower efficiency, ribbon-Si cells save on production costs due to a great reduction in waste because slicing silicon crystal into the thin wafers results in losses (of about 50%) of expensive pure silicon feedstock (Koch et al., 2003). Some of the manufacturing technologies of silicon ribbons are introduced into production, but their contribution to the Si-based solar energy is negligible (Fig. 1). The cost of ribbon-Si modules, as well as other types of silicon solar modules, remains quite high.

Many companies are developing solar cells that use lenses or/and mirrors to concentrate a large amount of sunlight onto a small area of photovoltaic material to generate electricity. This is the so-called **concentrated photovoltaics**. The main gain that is achieved through the involvement of concentrators is to save material. This, however, does not too reduce the cost of the device, because a number of factors lead to higher prices. (i) For the concentration of radiation, an optical system is necessarily required, which should maintain a solar cell in focus by the hardware when the sun moves across the sky. (ii) With a significant increase in the intensity of the radiation, the photocurrent also increases significantly, and the electrical losses rapidly increase due to voltage drop across the series-connected resistance of the bulk of the diode structure and contacts. (iii) For the removal of heat generated by irradiation that decreases the efficiency of photovoltaic conversion, it is necessary to use copper heatsinks. (iv) The requirements to quality of the solar cell used in the concentrator considerably increase. (v) Using concentrators, only direct beam of solar radiation is used, which leads to losing about 15% efficiency of solar module.

Nevertheless, today the efficiency of solar concentrators is higher compared to conventional modules, and this trend will intensify in the application of more efficient solar devices. In 2009, for example, the power produced in the world using solar energy concentration did not exceed 20-30 MW, which is ~ 0.1% of silicon power modules (Jager-Waldau, 2010). According to experts, concentrator market share can be expected to remain quite small although increasing by 25% to 35% per year (Von Roedern, 2006).

In general, in this protracted situation, experts and managers of some silicon PV companies have long come to conclusion that there would be limits to growing their wafer (ribbon) silicon business to beyond 1 GW per year by simply expanding further (Von Roedern, 2006). All of these companies are researching wafer-Si alternatives including the traditional *thin-film technologies* and are already offering such commercial thin-film modules.

Concluding this part of the analysis, one must agree, nevertheless, that wafer and ribbon silicon technology provides a fairly high rate of development of solar energy. According to the European Photovoltaic Industry Association (EPIA), the total installed PV capacity in the world has multiplied by a factor of 27, from 1.5 GW in 2000 to 39.5 GW in 2010 – a yearly growth rate of 40% (EPIA, 2011).

Undoubtedly, solar cells of all types on silicon wafers, representatives of the so-called *first generation photovoltaics*, will maintain their market position in the future. In hundreds of companies around the world, one can always invest (with minimal risk) and implement the silicon technology developed for microelectronics with some minor modifications (in contrast, manufacturers of thin-film solar modules had to develop their "own" manufacturing equipment). Monocrystalline and polycrystalline wafers, which are used in the semiconductor industry, can be made into efficient solar cells with full confidence. It is also important that silicon is very abundant, clean, nontoxic and very stable. However, due to limitations in production in large volumes of silicon for solar modules, which are both highly efficient and cost-effective, often-expressed projections for *desirable significant* increase in their contribution to the world energy system in the coming years are highly questionable.

3. Thin-film solar cells

Among the radical ways to reduce the cost of solar modules and to increase drastically the volume of their production is the transition to thin-film technology, the use of direct-gap semiconductors deposited on a cheap large-area substrate (glass, metal foil, plastic).

We start with the fact that the direct-gap semiconductor can absorb solar radiation with a thickness, which is much smaller than the thickness of the silicon wafer. This is illustrated by the results of calculations in Fig. 4 similar to those performed for the single-crystalline silicon shown in Fig. 2. Calculations were carried out for direct-gap semiconductors, which is already used as absorber layers of solar modules: a-Si, CdTe, $CuInSe_2$ and $CuGaSe_2$.

As expected, the absorptivity of solar radiation of direct-gap semiconductors in general is much stronger compared to crystalline silicon but the curves noticeably differ among themselves (in the references, the absorption curves for a-Si are somewhat different). Almost complete absorption of solar radiation by amorphous silicon (a-Si) in the $\lambda \leq \lambda_g = hc/E_g$ spectral range is observed at its thickness $d > 30\text{-}60$ µm, and 95% of the radiation is absorbed at a thickness of 2-6 µm (Fig. 4(a)). These data are inconsistent with the popular belief that in a-Si, as a direct-gap semiconductor, the *total* absorption of solar radiation occurs at a layer thickness of several microns. The total absorption of solar radiation in CdTe occurs if the thickness of the layer exceeds 20-30 µm, and 95% of the radiation is absorbed if the layer is thinner than ~ 1 µm. Absorptivities of the $CuInSe_2$ and $CuGaSe_2$ are even higher. Almost complete absorption of radiation in these materials takes place at a layer thickness of 3-4 µm, and 95% of the radiation is absorbed if the thickness of layer is only 0.4-0.5 µm (!).

Thus, the transition from crystalline silicon to direct-gap semiconductors leads to noncomparable less consumption of photoelectrically active material used in the solar cell.

High absorptivity of a semiconductor has important consequences with respect to other characteristics of the semiconductors used in solar cells. Since the direct-gap semiconductor can absorb solar radiation at its thickness much smaller than the thickness of the silicon wafer (ribbon), the requirements for chemical purity and crystalline perfection of the absorber layer in the solar cell became much weaker.

In fact, to collect photogenerated charge carriers, it is necessary to have a diffusion length of minority carriers in excess of the thickness of the absorbing layer. In the case of crystalline Si, the photogenerated carriers must be collected at a thickness of 1-2 hundred microns and 2 orders of magnitude smaller than in the case of CdTe, CIS or CGS. From this it follows that in the solar cell based on direct-gap semiconductor, the diffusion length L may be about two

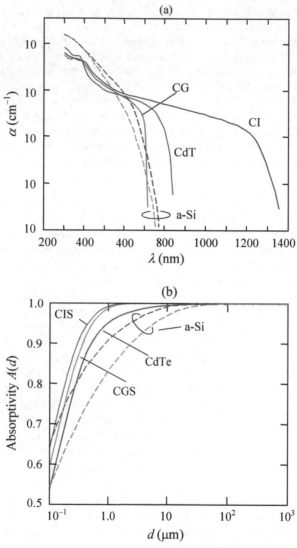

Fig. 4. Absorption curves (a) and dependence of absorptivity of AM1.5 solar radiation in the $h\nu \geq E_g$ spectral range on the absorber layer thickness d (b) for amorphous silicon (a-Si), cadmium telluride (CdTe), copper-indium diselenide (CIS) and copper gallium diselenide (CGS) (Han et al., 2007; Paulson et al., 2003; Gray et al., 1990; http://refractiveindex.info/a-Si).

orders of magnitude smaller, i.e. the carrier lifetime τ can be by 4(!) order shorter ($L \sim \tau^{1/2}$). Thus, the manufacture of thin film solar modules based on the direct-gap semiconductors does not require costly high purification and crystallinity of the material as it is needed in the production of modules based on crystalline, multicrystalline or ribbon silicon.

Thin-film technology has a number of other significant merits. While Si devices are manufactured from wafers or ribbons and then processed and assembled to form a modules, in thin-film technology many cells are simultaneously made and formed as a module. The layers of solar cells are deposited sequentially on moving substrates in a continuous highly automated production line (conveyor system) and, importantly, at temperatures not exceeding 200-650°C compared with 800-1450°C for the main processes of c-Si. This minimises handling and facilitates automation leading to the so-called *monolithic integration*.[4] Thin-film solar modules offer the lowest manufacturing costs, and are becoming more prevalent in the industry because allow to improve manufacturability of the production at significantly larger scales than for wafer or ribbon Si modules. Therefore, it is generally recognised that the contribution of thin-film technology in solar energy will be to grow from year to year faster. Many analysts believe *that it is only a matter of time before thin films would replace silicon wafer-based solar cells as the dominant photovoltaic technology*.

Unquestionable leaders in thin film technologies are solar cells on **amorphous silicon** (a-Si), **copper-indium-gallium diselenide** ($CuIn_xGa_{1-x}Se_2$) and **cadmium telluride** (CdTe), whose market share is expanding every year (Hegedus & Luque, 2011). The rest of the thin-film technologs are yet too immature to appear in the market but some of them is already reaching the level of industrial production. Below these technologies will first be briefly described, and a more detailed analysis of solar modules based on a-Si, CdTe and CIGS are allocated in separate subsections. Now the most successful non-Si based thin film PV technologies are representatives of the so-called *second generation photovoltaics* $CuIn_xGa_{1-x}Se_2$ and CdTe solar cells. Both of them have been manufactured in large scale and are commercialized.

(i) For a long time, intensive researches on own initiative and within different levels of government programs are carried out on developing **thin-film crystalline Si solar cells**. These devices are opposed to solar cells based on silicon wafers or ribbons because are made by depositing thin silicon layer on a foreign substrate. The thickness of such a layer can vary from a few tens of nanometers to tens of micrometers. Thin-film solar cells based on **crystalline silicon on glass substrate (CSG)** occupy a special place in these studies (Basore, 2006; Widenborg & Aberle, 2007). Such devices have the potential to reduce considerably the cost of manufacture of photovoltaic modules due to a significant thinning the absorbing layer and the use of cheap glass substrates. Of course, in a thin layer and a thick wafer of silicon, processes of collection of photogenerated carriers may substantially differ due to differences not only of layer thickness but also the structure of the material and its parameters such as the lifetime of carriers, their mobility and others. Because the mobility and lifetime of charge carriers in thin-film silicon layers are relatively low, the carrier diffusion lengths are generally lower than the penetration depths for the long-wavelength part of the solar spectrum and a narrow p-n junction cannot be employed in the thin-film silicon case. For this reason, one has to use p-i-n diode structures, where the photo-generation takes place in the i-layer and transport and collection are drift-assisted (Shah et al., 2006).

All the same, the thickness of Si layer is of great importance for other reasons. If in a typical case, the Si thickness is less than ~ 2 μm, an effective optical enhancement technique (light trapping) is necessary. Indeed, approximately only half of the solar radiation is absorbed in such layer, even when eliminating the reflection from the front surface of the solar cell (Fig. 3(b)).

---

[4] For example, First Solar manufactures the CdTe-based modules (120 cm × 60 cm, 70-80 W) on high throughput, automated lines from semiconductor deposition to final assembly and test – all in one continuous process. The whole flow, from a piece of glass to a completed solar module, takes less than 2.5 hours.

One effective way to obtain light trapping is to texture the supporting material (glass substrate) prior to the deposition of the Si film. To implement this idea, in particular, a glass aluminium-induced texturing (AIT) method was developed (Widenborg & Aberle, 2007). On the textured surface, silicon is deposited in amorphous form followed by solid-phase crystallisation and hydrogen passivation. An amorphous silicon is transformed into a polycrystalline layer after a special annealing at 400-600°C. As a result of the texture, light is transmitted obliquely into the Si film, significantly enhancing the optical pathlength and thus increasing the optical absorption. The effect is further enhanced by depositing a high-quality reflector onto the back surface. Best optical absorption is obtained if the texture and the back surface reflector are optimised such that the total internal reflection occurs both at the front and the rear surface of the Si film, enabling multiple passes of the light through the solar cell. There are other glass texturing methods compatible with producing poly-Si thin-film solar cells, for example, CSG Solar's glass bead method (Ji & Shi, 2002). Apart from the light trapping benefits, the textured substrate also reduces reflection losses at the front surface of the solar cell.

It should be noted again that the development of silicon solar cells on glass substrate is not limited to the problem of light trapping. Fabrication of these modules is also facing serious problems of differences in the thermal expansion coefficients of silicon and substrate, the influence of substrate material on the properties of a silicon thin layer at elevated temperatures and many others.

Despite the efforts of scientists and engineers for about 30 years, the stabilized efficiency of typical CSG devices still does not exceed 9-10%. Nevertheless, large-area CSG modules with such efficiency produce sufficient power to provide installers with a cost-effective alternative to conventional wafer or ribbon Si based products. Because of a low cost of production even with reduced efficiency, large-area CSG modules are attractive for some applications and are in production in factories having a capacity of tens of MW per year (Basore, 2006).

(ii) **Dye-sensitized-solar-cells** (DSSCs), invented by M. Grätzel and coworkers in 1991, are considered to be extremely promising because they are made of low-cost materials with simple inexpensive manufacturing procedures and can be engineered into flexible sheets (O'Regan & Grätzel, 1991; Grätzel, 2003; Chiba et al., 2006).

DSSCs are emerged as a truly new class of energy conversion devices. Mechanism of conversion of solar energy into electricity in these devices is quite peculiar. Unlike a traditional solar cell design, dye molecules in DSSC absorb sunlight, just as it occurs in nature (like the chlorophyll in green leaves). A porous layer of nanocrystalline oxide semiconductor (very often TiO_2) provides charge collection and charge separation, which occurs at the surfaces between the dye, semiconductor and electrolyte. In other words, the natural light harvest in photosynthesis is imitated in DSSC. DSSCs are representatives of the *third generation solar technology*. The dyes used in early solar cells were sensitive only in the short-wavelength region of the solar spectrum (UV and blue). Current DSSCs have much wider spectral response including the long-wavelength range of red and infrared radiation. It is necessary to note that DSSCs can work even in low-light conditions, i.e. under cloudy skies and non-direct sunlight collecting energy from the lights in the house.

The major disadvantage of the DSSC design is the use of the liquid electrolyte, which can freeze at low temperatures. Higher temperatures cause the liquid to expand, which causes problems sealing of the cell. DSSCs with liquid electrolyte can have the less long-term stability due to the volatility of the electrolyte contained organic solvent. Replacing the liquid electrolyte with a solid has been a field of research.

It should be noted that DSSCs can degrade when exposed to ultraviolet radiation. However, it is believed that DSSCs are still at the start of their development stage. Efficiency gain is possible and has recently started to be implemented. These include, in particular, the use of quantum dots for conversion of higher-energy photons into electricity, solid-state electrolytes for better temperature stability, and more.

Although the light-to-electricity conversion efficiency is less than in the best thin film cells, the DSSC price should be low enough to compete with fossil fuel electrical generation. This is a popular technology with some commercial impact forecast especially for some applications where mechanical flexibility is important. As already noted, energy conversion efficiencies achieved are low, however, it has improved quickly in the last few years. For some laboratory dye-sensitized-solar-cells, the conversion efficiency of 10.6% under standard AM 1.5 radiation conditions has been reached (Grätzel, 2004).

(iii) **Organic solar cells** attract the attention also by the simplicity of technology, leading to inexpensive, large-scale production. In such devices, the organic substances (polymers) are used as thin films of thickness ~ 100 nm. Unlike solar cells based on inorganic materials, the photogenerated electrons and holes in organic solar cells are separated not by an electric field of p-n junction. The first organic solar cells were composed of a single layer of photoactive material sandwiched between two electrodes of different work functions (Chamberlain, 1983). However, the separation of the photogenerated charge carriers was so inefficient that far below 1% power-conversion efficiency could be achieved. This was due to the fact that photon absorption in organic materials results in the production of a mobile excited state (exciton) rather than free electron-hole pairs in inorganic solar cells, and the exciton diffusion length in organic materials is only 5-15 nm (Haugeneder et al., 1999). Too short exciton diffusion length and low mobility of excitons are factors limiting the efficiency of organic solar cell, which is low in comparison with devices based on inorganic materials.

Over time, two dissimilar organic layers (bilayer) with specific properties began to be used in the organic solar cell (Tang, 1986). Electron-hole pair, which arose as a result of photon absorption, diffuses in the form of the exciton and is separated into a free electron and a hole at the interface between two materials. The effectiveness of ~ 7% reached in National Renewable Energy Laboratory, USA can be considered as one of best results for such kind of solar cells (1-2% for modules). However, instabilities against oxidation and reduction, recrystallization and temperature variations can lead to device degradation and lowering the performance over time. These problems are an area in which active research is taking place around the world. Organic photovoltaics have attracted much attention as a promising new thin-film PV technology for the future.

(iv) Of particular note are solar cells based on **III-V group semiconductors such as GaAs** and AlGaAs, GaInAs, GaInP, GaAsP alloys developed in many laboratories. These multi-junction cells consist of multiple thin films of different materials produced using metalorganic vapour phase epitaxy. Each type of semiconductor with a characteristic band gap absorbs radiation over a portion of the spectrum. The semiconductor band gaps are carefully chosen to generate electricity from as much of the solar energy as possible.

GaAs-based multi-junction devices were originally designed for special applications such as satellites and space exploration. To date they are the most efficient solar cells (higher than 41% under solar concentration and laboratory conditions), but the issue of large-scale use of GaAs-based solar cells in order to solve global energy problems is not posed (King, 2008; Guter at al., 2009).

Other solar cells have also been suggested, namely quantum dots, hot carrier cells, etc. However, they are currently studied at the cell-level and have a long way to be utilized in large-area PV modules.

3.1 Amorphous silicon

Amorphous silicon (a-Si) has been proposed as a material for solar cells in the mid 1970's and was the first material for commercial thin-film solar cells with all their attractiveness to reduce consumption of absorbing material, increase in area and downturn in price of modules.

It was discovered that the electrical properties of a-Si deposited from a glow discharge in silane (SiH_4) are considerably different from single-crystalline silicon (Deng & Schiff, 2003). When put into silane of a small amount of phosphine (PH_3) or boron (B_2H_6), electrical conduction of a-Si becomes n-type or p-type, respectively (Spear & Le Comber, 1975).

In 1976 Carlson and Wronski reported the creation of a-Si solar cells with efficiency of 2.4% using p-i-n structure deposited from a glow discharge in silane rather than evaporating silicon (Carlson & Wronski, 1976). The maximum efficiency of thin film amorphous silicon solar cells was estimated to be 14–15%.

a-Si is an allotropic form of silicon, in which there is no far order characteristic of a crystal. Due to this, some of a-Si atoms have nonsaturated bonding that appears as imperfection of the material and significantly affects its properties. The concentration of such defects is reduced by several orders due to the presence of hydrogen, which is always present in large quantities when obtained from the silane or at the surface treatment by hydrogen. The hydrogen atoms improve essentially the electronic properties of the plasma-deposited material. This material has generally been known as *amorphous hydrogenated silicon* (a-Si:H) and applied in the majority in practice.

Depending on the gas flow rate and other growth conditions, the optical band gap of a-Si:H varies, but typically ranges from 1.6 to 1.7 eV. Its absorption coefficient is much higher than that of mono-crystalline silicon (Fig. 4). As it has been noted, in the case of a-Si:H, the thickness of 2-6 µm (rather than 300 µm as in the case of c-Si) is sufficient for almost complete (95%) absorption of solar radiation in the $hv \geq E_g$ spectral range. It is also important that the technology of a-Si:H is relatively simple and inexpensive compared to the technologies for growing Si crystals. The low deposition temperature (< 300°C) and the application of the monolithic technique for a-Si:H module manufacturing were generally considered as key features to obtain low costs of the devices.

As in the past, the layers of a-Si:H can be deposited on large area (1 m² or more) usually by method termed as plasma enhanced chemical vapor deposition (PECVD) on glass coated with transparent conductive oxide (TCO) or on non-transparent substrates (stainless steel, polymer) at relatively low temperatures (a-Si:H can be also deposited roll-to-roll technology). Like the crystalline silicon, a-Si:H can be doped creating p-n junctions, which is widely used in other field of electronics, particularly in thin-film transistors (TFT).

This opens up the possibility of relatively easy to form the desired configuration of the active photodiode structure of the solar cell. To date, p-i-n junction is normally used in solar cells based on a-Si:H. The i-layer thickness is amount to several hundred nanometers, the thickness of frontal p-film, which is served as a "window" layer, is equal to ~ 20 nm, the back n-layer can be even thinner. It is believed that almost all electron-hole pairs are photogenerated in the i-layer, where they are separated by electric field of p-i-n structure. The output power of a-Si:H solar cell can has a *positive temperature coefficient*, i.e. at elevated ambient temperatures the efficiency is higher.

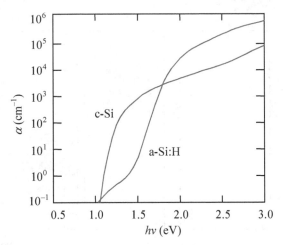

Fig. 5. Spectral dependence of absorption coefficient a in the crystalline (c-Si) and amorphous hydrogenated silicon (a-Si: H).

Quite common in commercial solar cells are the **multi-layer structures** based on amorphous silicon and silicon germanium alloys, when the p-i-n photodiode structures (subcells) with different band gap semiconductors are superimposed one layer on another. It seemed that the spectrum splitting tandem structure, a representative of the *third generation of solar cells*, opened the prospect of developing highly efficient and low-cost solar cells (Kuwano et al. 1982).

One of the tandem a-Si:H structures is shown in Fig. 6. This is the so-called superstrate design, when solar radiation enters through the transparent substrate such as glass or polymer (the substrate design with flexible stainless steel foil is also widely used). In the superstrate design, p- , i- and n-layers of a-Si:H are consistently applied on a glass plate coated with a transparent film (ITO, SnO_2). Over them the analogous layers of a-SiGe:H alloy are deposited in the discharge of silane CH_4 together with GeH_4. The frontal 0.5-µm thick layers of p-i-n structure absorb photons with energies larger than ~ 1.9 eV and transmits photons with energies lower than ~ 1.9 eV. The band gap of a-SiGe:H alloy is lower than that of a-Si, therefore the radiation that has passed through a-Si will be absorbed in the a-SiGe layers, where additional electron-hole pairs are generated and solar cell efficiency increases. Even greater effect can be achieved in a triple-junction structure a-Si/a-SiGe/a-SiGe. One of the record efficiency of such solar cell is 14.6% (13.0% stabilized efficiencies) (Yang et al,. 1997). At high content of Ge in $Si_{1-x}Ge_x$ alloy, optoelectronic properties of the material are deteriorated, therefore in multi-junction solar cells, the band gap of the amorphous $Si_{1-x}Ge_x$ layers cannot be less than 1.2-1.3 eV.

The results presented in Fig. 4, show that 10-15% of solar radiation power in the range $hv > E_g$ is not absorbed if the thickness of the a-Si layer is 1 µm. Therefore, to improve the power output, back reflector and substrate texturing can be used in a-Si solar cells. Apparently, the light trapping occurs for weakly absorbed light. It was shown that using geometries maximizing enhancement effects, the short circuit current in amorphous silicon solar cell (< 1 µm thick) increases by several mA/cm² (Deckman et al., 1983).

The use of multi-junction solar cells is successful because there is no need for lattice matching of materials, as is required for crystalline heterojunctions.

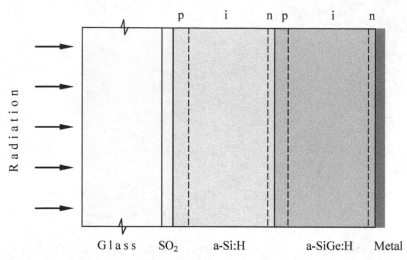

Fig. 6. Tandem solar cell based on a-Si:H and a-SiGe:H (Deng & Schiff, 2003).

Solar cells based on a-Si:H are much cheaper than those produced on silicon wafers or ribbons, but their efficiency in operation under illumination becomes lower during the first few hundred hours and then the degradation process is slowed down considerably (Fig. 7). The degradation of multiple-junction and single-junction solar cells is usually in the range of 10-12 and 20-40%, respectively (20-30% for commercial devices).

Degradation of a-Si:H solar cells, called the Staebler–Wronski effect, is a *fundamental* a-Si property. The same degradation is observed in solar cells from different manufacturers and with different initial efficiencies (Von Roedern et al., 1995). It has been established that stabilization of the degradation occurs at levels that depend on the operating conditions, as well as on the operating history of the modules. After annealing for several minutes at 130-150°C, the solar cell properties can be restored. The positive effect of annealing can also occur

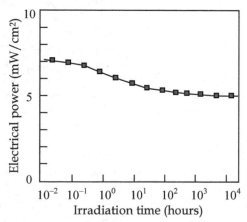

Fig. 7. Decline in power output of solar cell based on a-Si:H in the initial period of irradiation 100mV/sm² (Staebler–Wronski effect) (Deng & Schiff, 2003).

as a result of *seasonal* temperature variations therefore the efficiency of a-Si:H modules in the summer is greater than in winter.

Degradation of a-Si:H solar cells is mostly caused by photostimulated formation of defects (dangling bonds) that act as recombination centers (Staebler & Wronski, 1977). The effect can not be explained by a single degradation mechanism. At least two mechanisms have to be involved: a fast one that can be annealed at typical module operating temperatures, and a slow one that does not recover measurably when annealing temperatures are limited to values below 70°C (Von Roedern et al., 1995; Von Roedern & del Cueto, 2000).

It is important from a practical point of view that a-Si:H reaches a "stabilized" state after extended irradiation. The stabilized a-Si:H arrays show less than 1% degradation per year, which is about the same rate at which crystalline silicon loses power over time. Therefore, still 20 years ago, it was recommended that all a-Si solar cells and modules performances should be reported after stabilization under standard conditions for 1000 hours at 50°C (Von Roedern & del Cueto, 2000).

Along with amorphous silicon, during the same 30 years, the intensive investigations of the possibility of using the glow discharge method for producing **thin-film crystalline silicon** were also carried out (Faraji et al., 1992; Meier et al., 1994). In 1994, using this method thin-film p-i-n solar cells based on hydrogenated **microcrystalline silicon (μc-Si)** have been prepared at substrate temperatures as low as ~ 200°C (Meier et al., 1994). Compared with a-Si:H solar cells, an enhanced absorption in the near-infrared (up to 1.1 eV) and an efficiency of 4.6% were obtained. First light-soaking experiments indicate also no degradation for such μc-Si:H cells. Years later, other groups started research activities utilizing this material and naming the microcrystalline silicon as **"nanocrystalline" (nc-Si)** or **"policrystalline" silicon (poly-Si)**.

Microcrystalline silicon (μc-Si, nc-Si,poly-Si) has small grains of crystalline silicon (< 50 nm) within the amorphous phase. This is in contrast to multi-crystalline silicon, which consists solely of crystalline silicon grains separated by grain boundaries. Microcrystalline silicon has a number of useful advantages over a-Si, because it can have a higher electron mobility due to the presence of the silicon crystallites. It also shows increased absorption in the red and infrared regions, which makes μc-Si as an important material for use in solar cells. Another important advantages of μc-Si is that it has improved stability over a-Si, one of the reasons being because of its lower hydrogen concentration. There is also the advantage over poly-Si, since the μc-Si can be deposited using conventional low temperature a-Si deposition techniques, such as plasma enhanced chemical vapor deposition.

Special place in the thin-film photovoltaics is the so-called **micromorph solar cells**, which are closely related to the amorphous silicon. The term "micromorph" is used for stacked tandem thin-film solar cells consisting of an a-Si p-i-n junction as the top component, which absorbs the "blue" light, and a μc-Si p-i-n junction as the bottom component, which absorbs the "red" and near-infrared light in the solar spectrum (Fig. 8). This artificial word has first been mentioned by Meier in the late 1990's (Meier et al., 1995; Meier et al., 1998).

A double-junction or tandem solar cell consisting of a microcrystalline silicon solar cell (E_g = 1.1 eV) and an amorphous silicon cell ($E_g \approx 1.75$ eV) corresponds almost to the theoretically optimal band gap combination. Using μc-Si:H as the narrow band gap cell instead of a-SiGe cell yields the higher efficiency in the long-wavelength region. At the same time, the stable μc-Si:H bottom cell contributes to usually a better stability of the entire micromorph tandem cell under light-soaking. The stabilized efficiency of such double-junctions solar cells of 11-12% was achieved (Meier et al., 1998).

It was also suggested that the micromorph solar cells open new perspectives for low-temperature thin-film crystalline silicon technology and even has the potential to become for the next third generation of thin-film solar cells (Keppner et al., 1999). However, due to the effect of the indirect band, the μc-Si component of solar cells requires much thicker i-layers to absorb the sunlight (1.5 to 2 μm compared to 0.2 μm thickness of a-SiGe:H absorber). At the same time, the deposition rate for μc-Si:H material is significantly lower compared to that for a-SiGe, so that a much longer time is needed to deposit a thicker μc-Si layer than what is needed for an a-SiGe structure. In addition, advanced light enhancement schemes need to be used because μc-Si has a lower absorption coefficient. Finally, μc-Si solar cell has a lower open-circuit voltage compared to a-SiGe:H cell.

Nevertheless, one should tell again that micromorph tandem solar cells consisting of a microcrystalline silicon bottom cell and an amorphous silicon top cell are considered as one of the promising thin-film silicon solar-cell devices. The promise lies in the hope of simultaneously achieving high conversion efficiencies at relatively low manufacturing costs.

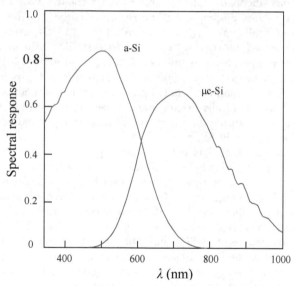

Fig. 8. Spectral response of a double-stacked micromorph tandem solar cell (Keppner et al., 1999).

Concluding the section, one must admit the considerable progress for more than 30 years in improving the efficiency of a-Si and micromorph solar cells. However, despite the persistent efforts of researchers and engineers in various laboratories in many countries, under initiatory researches and government programs, stabilized efficiency of large-area a-Si:H solar modules lies within 8 to 10%. So far, stabilized efficiencies of about 11-12% can be obtained with micromorph solar cells. A number of technological methods allow the efficiency just somewhat to increase using the above alloys of Si-Ge and special multiple-junction structures including the use of μc-Si:H and others.[5] But the cost of such multiple-

---

[5] One of the highest stabilized cell efficiency for a laboratory triple-junction structure is ~ 13.0% (Deng & Schiff, 2003).

junction solar cells always increases markedly. In addition, faster deposition processes need to be developed that is necessary for low-cost and high-throughput manufacturing.

Thus, in the field of amorphous and micromorph silicon photovoltaics, it is succeeded in realizing only part of the benefits of thin-film solar cells, and its share in the global solar energy is quite small (Fig. 1). The use of a-Si:H and micromorph solar cells is limited preferably to areas where low cost is more important than the efficiency of photoelectric conversion (such as consumer electronics). Semitransparent modules are also used as architectural elements or windows and skylights. This is the so-called building-integrated photovoltaics (BIPV), where large building envelope areas can be covered with the PV modules available at the lowest prices per square meter. The advantage of BIPV is also that the initial cost can be offset by reducing the amount spent on building materials. This trend becomes noticeable segment of the photovoltaic industry.

3.2 Copper-indium-gallium diselenide

$CuInSe_2$ (CIS) and $CuGaSe_2$ (CGS), compound semiconductors of elements from groups I, III and VI in the periodic table, are typical representatives of a broad class of substances with different properties, so-called chalcopyrite.[6] $CuInSe_2$ and $CuGaSe_2$ form $CuIn_xGa_{1-x}Se_2$ alloy (CIGS) in any ratio of components. Importantly, varying the ratio of $CuInSe_2$ and $CuGaSe_2$, there is a slight change in the material parameters except the band gap. The possibility of regulating the band gap semiconductor in the range of 1.0-1.04 eV for $CuInSe_2$ to 1.68 eV for $CuGaSe_2$ are doubtless advantages of $CuIn_xGa_{1-x}Se_2$ (Birkmire, 2008).[7] The dependence of the band gap of $CuIn_xGa_{1-x}Se_2$ on x is described by the formula (Alonso, 2002):

$$E_g(x) = 1.010 + 0.626x - 0.167x(1-x). \tag{2}$$

For a long time, $CuInSe_2$ and $CuIn_xGa_{1-x}Se_2$ have been considered as promising materials for high-performance thin-film solar cells and fabrication of monolithically interconnected modules intended for cost-effective power generation (Shafarman & Stolt, 2003).

Parameters of $CuIn_xGa_{1-x}Se_2$ can be fitted as optimal for the photoelectric conversion. The fundamental absorption edge is well described by expression $a \sim (hv - E_g)^{1/2}/hv$ as for a typical direct band gap semiconductor. When the photon energy hv exceeds the band gap E_g, the absorption coefficient of material from any content of Ga quickly exceeds values $\sim 10^4$ cm^{-1} so that the absorptivity of the material turns out to be the largest among all thin film (Fig. 4). This ensures effective absorption of solar radiation by $CuIn_xGa_{1-x}Se_2$ layer of micron-level thick – an important factor in reducing the cost of production. Efficiency of $CuInSe_2$ solar cell is within 12-15%, and in the case of $CuIn_xGa_{1-x}Se_2$ a record value of $\sim$ 20% is reached among all types of thin-film solar cells (Repins, et al., 2008; Green et al, 2011). Widening the band gap of $CuIn_xGa_{1-x}Se_2$ leads to an increase in open circuit voltage while reducing the absorptivity of the material and hence decreasing the short circuit current. Theoretical estimation shows that the maximum efficiency of solar cells should be observed when E_g = 1.4-1.5 eV, when the atomic ratio for Ga in CIGS is $\sim$ 0.7, but according to the experimental data, this occurs when $E_g \approx 1.15$ eV that corresponds to Ga/(Ga+In) ≈ 0.3.

---

[6] Chalcopyrite (copper pyrite) is a mineral $CuFeS_2$, which has a tetragonal crystal structure.
[7] Sometimes part of Se atoms substitute for S.

Structural, electrical and optical properties of $CuIn_xGa_{1-x}Se_2$ are sensitive to deviations from the stoichiometric composition, native defects and grain sizes in the film. Because of the native defects (mainly In vacancies and Cu atoms on In sites), the conductivity of CIGS is p-type. It can be controlled by varying the Cu/In ratio during growth of the material. Stoichiometric and copper-enriched material has a p-type conductivity and grain sizes of 1-2 µm, while the material indium-enriched has n-type conductivity and smaller grains. $CuIn_xGa_{1-x}Se_2$ within the solar cell contains a large amount of Na (~ 0.1%), which are predominantly found at the grain boundaries rather than in the bulk of the grains and can improve solar cell performance. Depending on the pressure of selenium, the type conductivity of material as a result of annealing can be converted from p-type to n-type – or vice versa. Charge carrier density can vary widely from 10^{14} cm^{-3} to 10^{19} cm^{-3}. In general, the desirable conductivity and carrier concentration can be relatively easy to obtain without special doping, but only the manipulating technological conditions of $CuIn_xGa_{1-x}Se_2$ deposition.

Literature data on the charge carrier mobility in thin-film $CuIn_xGa_{1-x}Se_2$ are quite divergent. The highest hole mobility is fixed as 200 cm^2/V·s at 10^{17} cm^{-3} hole concentration. It is likely that the conductivity across grain boundaries in this case plays a significant role. In $CuIn_xGa_{1-x}Se_2$ single crystals, mobility of holes lies within the 15-150 cm^2/V·s range for holes and in the 90-900 cm^2/V·s range for electrons (Neumann & Tomlinson, 1990; Schroeder & Rockett, 1997).

The first photovoltaic structures based on $CuInSe_2$ with efficiency of ~ 12% were established back in 1970's by evaporating n-CdS onto p-$CuInSe_2$ single crystal (Wagner et al., 1974; Shay et al., 1975). Shortly thin-film CdS/$CuInSe_2$ solar cells were fabricated with efficiency 4-5% (Kazmerski et al., 1976), interest to which became stable after the Boeing company had reached 9.4% efficiency in 1981 (Mickelson & Chen, 1981; Shafarman & Stolt, 2003). Such solar cells were produced by simultaneous thermal evaporation of Cu, In and Se from separate sources on heated ceramic substrates coated with thin layer of Mo (*thermal multi-source co-evaporation process*). Later the method of simultaneous deposition Cu, In, Ga and Se become widely used to create $CuIn_xGa_{1-x}Se_2$ layers. Chemical composition of material is determined by temperatures of sources: 1300-1400° C for Cu, 1000-1100°C for In, 1150-1250°C for Ga and 300-350°C for Se. The main advantage of this technology is its flexibility; the main problem is the need for careful control of flow of Cu, In, Ga and Se, without which it is impossible to have adequate reproducibility characteristics of the film. In this regard, attractive is the so-called *two-step process*, that is, the deposition of Cu, In and Ga on substrates at a low temperature with subsequent a reactive heat-treatment of Cu-In-Ga films in a hydrogen-selenium (H_2Se) atmosphere at temperatures above ~ 630°C (Chu et al., 1984). Application of Cu, In and Ga can be achieved by various methods at low temperatures, among them is ion sputtering, electrochemical deposition and other methods that are easier to implement in mass production. Selenization can be conducted at atmospheric pressure at relatively low temperatures of 400-500°C. The main problem of this technology is the complexity of controlling the chemical composition of material as well as high toxicity of H_2Se.

To date, the co-deposition of copper, indium, gallium and selenium as well as selenization remain the main methods in the manufacture of CIGS solar cells.

In the first thin-film $CuInSe_2$ solar cells, heterojunction was made by deposition of CdS on $CuInSe_2$ thin film, which also served as front transparent electrode (Mickelson & Chen, 1981). Characteristics of solar cell are improved if on $CuInSe_2$ (or $CuIn_xGa_{1-x}Se_2$) first to deposit undoped CdS, and then low-resistive CdS doped with In or Ga (pre-inflicted undoped CdS layer is called the "buffer" layer). Due to a relatively narrow band gap (2.42 eV), CdS absorbs solar radiation with a wavelengths λ < 520 nm, without giving any contribution to the photovoltaic efficiency. Absorption losses in the CdS layer can be reduced by increasing the band gap, alloying with ZnS (CdZnS) that results in some increase in the efficiency of the device. Its further increase is achieved by thinning CdS layer to 50 nm or even 30 nm followed by deposition of conductive ZnO layer, which is much more transparent in the whole spectral region (Jordan, 1993; Nakada, T. & Mise, 2001). The best results are achieved when ZnO is deposited in two steps, first a high-resistance ZnO layer and then a doped high-conductivity ZnO layer. Often, ZnO films are deposited by magnetron sputtering from $ZnO:Al_2O_3$ targets or by reactive sputtering, which requires special precision control technology regime. For high-efficiency cells the TCO deposition temperature should be lower than 150°C in order to avoid the detrimental interdiffusion across CdS/CIGS interface (Romeo et al., 2004).

Usually, $Cu(In,Ga)Se_2$ solar cells are grown in a substrate configuration which provides favorable process conditions and material compatibility. Structure of a typical solar cell is shown in Fig. 9. To reduce the reflection losses at the front surface of ZnO, an anti-refection MgF_2 coating with thickness of ~ 100 nm is also practised. The substrate configuration of solar cell requires an additional encapsulation layer and/or glass to protect the cell surface. In modules with cover glasses, to use any anti-refection coating is not practical.

R a d i a t i o n

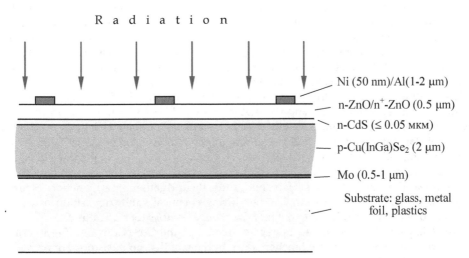

Ni (50 nm)/Al(1-2 µm)

n-ZnO/n$^+$-ZnO (0.5 µm)

n-CdS ($\leq$ 0.05 мкм)

p-$Cu(InGa)Se_2$ (2 µm)

Mo (0.5-1 µm)

Substrate: glass, metal foil, plastics

Fig. 9. Schematic cross section of a typical $Cu(In,Ga)Se_2$ solar module.

CdS layer is made by chemical precipitation from an aqueous alkali salt solution of cadmium ($CdCl_2$, $CdSO_4$, CdI2, $Cd(CH3COO)_2$), ammonia (NH_3) and thiourea ($Sc(NH_2)_2$ in molar ratio, for example, 1.4:1:0.1 (*chemical bath deposition*). Pseudo-epitaxial deposition of CdS dense films is carried out by immersing the sample in electrolyte for several minutes at temperatures from 60 to 80°C or at room temperature followed by heating electrolyte to the same temperature. The pseudo-epitaxial character of deposition is promoted, firstly, by small (~ 0.6%) difference of $CuInSe_2$ and CdS lattice spacing, which, however, increases with increasing Ga content in $CuInxGa_{1-x}Se_2$ (to ~ 2% at x = Ga/(Ga+In) = 0.5), and, secondly, by the cleansing effect of electrolyte as a surface etchant of $CuIn_xGa_{1-x}Se_2$ (ammonia removes oxides on the surface). Depending on the conditions of deposition, the film may have hexagonal, cubic or a mixed structure with crystallite sizes of several tens of nanometers. Typically, film is somewhat non-stoichiometric composition (with an excess of Cd) and contains impurities O, H, C, N that can become apparent in a noticeable narrowing of the band gap. It is believed that the Cd in Cu(InGa)Se$_2$ modules can be handled safely, both with respect to environmental concerns and hazards during manufacturing (Shafarman & Stolt, 2003).

At relatively low temperature of deposition, the mutual penetration (migration) of elements at the CdS/$CuIn_xGa_{1-x}Se_2$ interface takes place to a depth of 10 nm (Cd replace Cu). It should be noted that vacuum deposition of CdS, used in solar cells on single crystals $CuIn_xGa_{1-x}Se_2$, is not suitable for thin film structures and does not allow to obtain the dense film of necessary small thickness and requires too high deposition temperature (150-200°C). Deposition of CdS by ion sputtering gives better results, but still inferior to chemical vapor deposition.

Metal contacts in the form of narrow strips to the front surface of Cu(In,Ga)Se$_2$ device is made in two steps: first a thin layer of Ni (several tens of nanometers), and then Al layer with thickness of several microns. Purpose of a thin layer is to prevent the formation of oxidation layer.

As substrate for $CuIn_xGa_{1-x}Se_2$ solar cells, the window soda-lime-silica glass containing 13-14% Na$_2$O can be used. The coefficients of linear expansion of this glass and $CuIn_xGa_{1-x}Se_2$ are quite close ($9{\times}10^{-6}$ K^{-1}) in contrast to borosilicate glass, for which the coefficient of linear expansion is about half. Glass is the most commonly used substrate, but significant efforts have been made to develop flexible solar cells on polyimide and metal foils providing less weight and flexible solar modules. Highest efficiencies of 12.8% and 17.6% have been reported on polyimide and metal foils, respectively (Tiwari etal., 1999; Tuttle et al., 2000). Cu(In,Ga)Se$_2$ modules have shown stable performance for prolonged operation in field tests.

As already mentioned, it is believed that the p-n junction is formed between p-$CuIn_xGa_{1-x}Se_2$ and n-ZnO, "ideal" material that serves as a "window" of solar cell (ZnO has band gap of 3.2 eV, high electrical conductivity and thermal stability). However, a thin underlayer CdS (~ 0.05 nm) affect a strong influence on the characteristics of solar cell by controlling the density of states at the interface and preventing unwanted diffusion of Cu, In, Se in ZnO. Somewhat simplified energy diagram of solar cell based on $CuIn_xGa_{1-x}Se_2$ is shown in Fig. 10.

Band discontinuity ΔE_c = 0.3 eV at the CdS/$CuIn_xGa_{1-x}Se_2$ interface causes considerable band bending near the $CuIn_xGa_{1-x}Se_2$ surface, and, thus, the formation of p-n junction (Schmid et al., 1993). Diffusion of Cd in $CuIn_xGa_{1-x}Se_2$ during chemical vapor deposition of CdS also promotes this resulting in forming p-n *homojunction near surface* of $CuIn_xGa_{1-x}Se_2$.

Marginal impact of losses caused by recombination at the CdS/$CuIn_xGa_{1-x}Se_2$ interface is explained by the creation of p-n junction, despite the fact that no measures are preventable to level the lattice difference and defects on the surface which is in the air before deposition of CdS.

As always, the short-circuit current of $CuIn_xGa_{1-x}Se_2$ solar cell is the integral of the product of the external quantum efficiency and the spectral density of solar radiation power. QE_{ext}, which, in turn, is determined primarily by the processes of photoelectric conversion in the $CuIn_xGa_{1-x}Se_2$ absorber layer, i.e. by the internal quantum yield of the device QE_{int}.

It is believed that the solar cell can neglect recombination losses at the CdS/Cu(In,Ga)Se$_2$ interface and in the space-charge region and then one can write (Fahrenbruch A. & Bube, 1983):

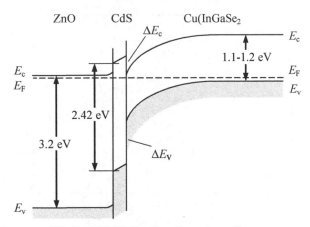

Fig. 10. Energy diagram of ZnO/CdS/CuIn$_x$Ga$_{1-x}$Se$_2$ solar cell.

$$QE_{int} = 1 - \frac{\exp(-\alpha W)}{1 + \alpha L_n},$$ (3)

where a is the light absorption coefficient, and W is the space-charge region width.

Besides QE_{int}, external quantum efficiency is also controlled by the above-mentioned reflection at the front surface of the device, reflection at all other interfaces, the band gap of CuIn$_x$Ga$_{1-x}$Se$_2$ and the transmittances of CdS and ZnO window layers.

Fig. 11 shows the measured spectral distribution of quantum efficiency of solar cells based on CuIn$_x$Ga$_{1-x}$Se$_2$ with different composition x = 0, 0.24 and 0.61, and hence with different band gap of semiconductor E_g = 1.02, 1.16 and 1.40 eV, respectively.

Another important characteristic of CuIn$_x$Ga$_{1-x}$Se$_2$ solar cell, the open-circuit voltage, is determined by the charge transport mechanism in the heterostructure. Neglecting recombination at the interface of CdS-CuIn$_x$Ga$_{1-x}$Se$_2$, the current-voltage characteristics of solar cells can be presented in the form

$$J = J_d - J_{ph} = J_o \exp\left[\frac{q}{nkT}(V - R_s J)\right] + GV - J_{ph}$$ (4)

where J_d is the dark current density, J_{ph} is the photocurrent density, n is the ideality factor, R_s is the series resistance, and G is the shunt conductivity.

The experimental curves are often described by Eq. (4) at n = 1.5 ± 0.3 that leads to the conclusion that the dominant charge transfer mechanism is recombination in the space charge region. If recombination level is located near mid-gap, $n \approx 2$, and in case of shallow level $n \approx 1$. In real CuIn$_x$Ga$_{1-x}$Se$_2$, the levels in the band gap are distributed quasi-continuously.

If the minority carrier diffusion length is short, the losses caused by recombination at the rear surface of CuIn$_x$Ga$_{1-x}$Se$_2$ is also excluded. In the best solar cells the electron lifetime is 10^{-8}-10^{-7} s (Nishitani et al., 1997; Ohnesorge et al., 1998). When describing transport properties CuIn$_x$Ga$_{1-x}$Se$_2$, it can to be acceptable that grain boundaries do not play any noticeable role since the absorber layer has a *columnar structure* and the measured current does not cross the grain boundaries. As notes, solar cells have the highest photovoltaic efficiency if x = Ga/(In + Ga) $\approx$ 0.3, i.e., $E_g \approx$ 1.15 eV. Under AM1.5 global radiation, the

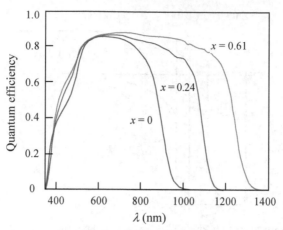

Fig. 11. Spectral distribution of quantum efficiency of $CuIn_xGa_{1-x}Se_2$ solar cells with x = 0, 0.24 and 0.61 (Shafarman & Stolt, 2003).

highest value of short-circuit current density J_{sc} = 35.2 mA/sm² is observed for solar cells with E_g = 1.12 eV (Contreras et al., 1999). If short-circuit current decreases with increasing Ga content, the open-circuit voltage V_{oc} increases. With increasing temperature V_{oc} markedly reduces. For E_g = 1.16 eV, for example, V_{oc} reduces from ~ 0.75 V at 220 K to ~ 0.55 V at 320 K. Introduction of Ga in $CuInSe_2$ compound attracts of professionals by the fact that it reduces the cost of In, which is widely used in LCD monitors, computers, TV screens and mobile phones. Therefore there is an attempt to reduce the content of In in $CuIn_xGa_{1-x}Se_2$ solar cells up to 5-10%, even slightly losing the photovoltaic conversion efficiency.

The efficiencies of laboratory $CuIn_xGa_{1-x}Se_2$ solar cells and modules of large area are significantly different. The reason is that the production of modules requires the introduction of technology different qualitatively from that used in the traditional semiconductor electronics, and a significant lack of deep scientific basis of applied materials. As a result of research, aimed to reducing the cost of $CuIn_xGa_{1-x}Se_2$ solar modules (which were originally more expensive compared to devices on amorphous silicon), Würth Solar (Germany) and Shell Solar Industries (USA) developed the first commercial $CuIn_xGa_{1-x}Se_2$ solar modules and initiated their large-scale production, which began in 2006 in Germany.

In the production of such modules are also engaged other companies in a number of countries, among them Zentrum für Sonnenenergie- und Wasserstoff-Forschung – ZSW (Germany), Energy Photovoltaics, Inc. and International Solar Electric Technology (USA), Angstrom Solar Centre (Sweden), Showa Shell and Matsushita (Japan) and others. Technology for production of solar modules on flexible substrates involving «roll-to-roll» technology was developed by Global Solar Energy (USA, Germany).

$CuIn_xGa_{1-x}Se_2$-based photovoltaics, along with other thin-film PV devices, continue to attract an interest first and foremost because of their potential to be manufactured at a lower cost than Si wafer or ribbon based modules. To reach their potential for large-scale power generation with higher throughput, yield, and performance of products, there is a need for continued improvement in the fundamental science, deposition equipment and processes based on well-developed models. Note also that the scarce supply of In may make it difficult to implement CIGS technology on a large scale.

3.3 Cadmium telluride

Cadmium telluride (CdTe) is a semiconductor with the band gap of 1.47-1.48 eV (290-300 K), optimal for solar cells. As a-Si, CIS and CIGS, CdTe is a direct-gap semiconductor, so that the thickness of only a few microns is sufficient for almost complete absorption of solar radiation (97-98%) with photon energy $hv > E_g$ (Fig. 4). As the temperature increases the efficiency of CdTe solar cell is reduced less than with silicon devices, which is important, given the work of solar modules in high-power irradiation. Compared to other thin-film materials, technology of CdTe solar modules is simpler and more suitable for large-scale production.

Solar cells based on CdTe have a rather long history. Back in 1956, Loferski theoretically grounded the use of InP, GaAs and CdTe in solar cells as semiconductors with a higher efficiency of photoelectric conversion compared with CdS, Se, AlSb and Si (Loferski, 1956). However, the efficiency of laboratory samples of solar cells with p-n junctions in monocrystalline CdTe, was only ~ 2% in 1959, has exceeded 7% only in 20 years and about 10% later (Minilya-Arroyo et al, 1979; Cohen-Solal et al., 1982). The reason for low efficiency of these devices were great losses caused by surface recombination and technological difficulties of p-n junction formation with a thin front layer. Therefore, further efforts were aimed at finding suitable *heterostructures*, the first of which was p-Cu$_2$Te/n-CdTe junction with efficiency of about 7%, that was proved too unstable through the diffusion of copper. It was investigated other materials used as heteropartners of n-type conductivity with wider band gap compared with CdTe: ITO, In$_2$O$_3$, ZnO performed the function of "window" through which light is introduced in the photovoltaic active layer of absorbing CdTe.

In 1964, the first heterojunctions obtained by spraying a thin layer of n-CdS on the surface of p-CdTe single crystal were described (Muller & Zuleeg, 1964). The first *thin-film* CdTe/CdS/SnO$_2$/glass structures that became the prototype of modern solar cells, was established in Physical-Technical Institute, Tashkent, Uzbekistan in 1969 (Adirovich et al., 1969). Over the years it became clear that the CdS/CdTe heterostucture has a real prospect of the introduction into mass production of solar modules, despite the relatively narrow band gap of CdS as a "window" layer. The crystal of CdTe adopts the wurtzite crystal structure, but in most deposited CdTe films, hexagonally packed alternating Cd and Te layers tend to lie in the plane of the substrate, leading to columnar growth of crystallites. At high temperature, CdTe grows stoichiometrically in thin-film form as natively p-doped semiconductor; no additional doping has to be introduced. Nevertheless, the cells are typically "activated" by using the influence of CdCl$_2$ at elevated temperatures (~ 400°C) that improves the crystallinity of the material.

In the early 21st century it has been succeeded to achieve a compromise between the two main criteria acceptable for manufacturing CdTe solar modules – sufficient photoelectric conversion efficiency and cheapness of production (Bonnet, 2003). This was possible thanks to the development of a number of relatively simple and properly controlled method of applying large area of CdTe and CdS thin layers that is easy to implement in large-scale production: close-space sublimation, vapor transport deposition, electrodeposition, chemical bath deposition, sputter deposition, screen printing. Obstruction caused by considerable differences of crystal lattice parameters of CdTe and CdS (~ 5%), largely overcome by straightforward thermal treatment of the produced CdTe/CdS structure. It is believed that this is accompanied by a mutual substitution of S and Te atoms and formation an intermediate CdTe$_{1-x}$S$_x$ layer with reduced density of states at the interface of CdTe and CdS, which may adversely affect the efficiency of solar cell. Simple methods of production and formation of barrier structures, that do not require complex and expensive equipment, are an important advantage of the solar cell technology based on CdTe.

When producing solar cells, CdS and CdTe layers are usually applied on a soda-lime glass superstrate (~ 3 mm thick), covered with a transparent electrically conductive oxide layer (TCO), e.g., F-doped SnO_2 (SnO_2:F) or ITO (In_2O_3 + SnO_2) (Fig. 12) (Bonnet, 2003).[8] They are often used in combination with a thin (high-resistivity) SnO_x sublayer between the TCO and the CdS window layer, which prevents possible shunts through pinholes in the CdS and facilitates the use of a thinner CdS layer for reducing photon absorption losses for wavelengths shorter than 500 nm (Bonnet, 2002). At the final stage, after deposition of the back electrodes, solar cells are covered by another glass using the sealing material (etylenvinil acetate, EVA), which provides durability and stability of the devices within 25-35 years.

Processes of photoelectric conversion in thin-film CdS/CdTe structure are amenable to *mathematical description*. This is of practical importance because it allows to investigate the dependence of the efficiency of solar cells on the parameters of the materials and the barrier structure as well as to formulate recommendations for the technology. These parameters are, primarily, (i) the width of the space-charge region, (ii) the lifetime of minority carriers, (iii) their diffusion length, (iv) the recombination velocity at the front and back surfaces of the CdTe absorber layer, (v) its thickness.

R a d i a t i o n

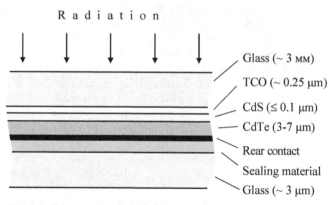

Glass (~ 3 MM)

TCO (~ 0.25 μm)

CdS (≤ 0.1 μm)

CdTe (3-7 μm)

Rear contact

Sealing material

Glass (~ 3 μm)

Fig. 12. Cross-section of thin film solar cell CdS/CdTe.

One of the main characteristics of a solar cell is the *spectral distribution of quantum efficiency* (spectral response), which is ultimately determined the short-circuit current density of the CdS/CdTe heterostructure.

It is known that in CdS/CdTe solar cells only the CdTe layer contributes to the light-to-electric energy conversion, while the CdS "window" layer only absorbs light in the range λ < 500-520 nm thereby reducing the photocurrent. Therefore in numerous papers a band bending (and hence a depletion layer) in CdS is not depicted on the energy diagram (see, for example, Birkmire & Eser, 1997; Fritsche et al., 2001; Goetzberger et al, 2003), i.e. the

---

[8] The CdTe solar cells can be produced in both substrate and superstrate configurations, but the latter is preferable. The substrate can be a low-cost soda-lime glass for growth process temperatures below 550°C, or alkali-free glass for high-temperature processes (550–600°C) (Romeo et al., 2004).

depletion layer of the CdS/CdTe diode structure is virtually located in the p-CdTe layer (Fig. 13). This is identical to the case of an asymmetric abrupt p-n junction or a Schottky diode, i.e. the potential energy $\varphi(x,V)$ and the space-charge region width W in the CdS/CdTe heterojunction can be expressed as (Sze, 1981):

$$\varphi(x,V) = (\varphi_o - qV)\left(1 - \frac{x}{W}\right)^2,$$ (5)

$$W = \sqrt{\frac{2\varepsilon\varepsilon_o(\varphi_o - qV)}{q^2(N_a - N_d)}},$$ (6)

where ε_o is the electric constant, ε is the relative dielectric constant of the semiconductor, $\varphi_o = qV_{bi}$ is the barrier height at the semiconductor side (V_{bi} is the built-in potential), V is the applied voltage, and $N_a - N_d$ is the uncompensated acceptor concentration in the CdTe layer. The *internal* photoelectric quantum efficiency η_{int} can be found from the continuity equation with the boundary conditions. The exact solution of this equation taking into account the *drift* and *diffusion components* as well as surface recombination at the interfaces leads to rather cumbersome and non-visual expressions (Lavagna et al., 1977). However, in view of the real CdS/CdTe thin-film structure, the expression for the *drift* component of the quantum efficiency can be significantly simplified (Kosyachenko et al., 2009):

$$\eta_{drift} = \frac{1 + \dfrac{S}{D_n}\left(\alpha + \dfrac{2}{W}\dfrac{\varphi_o - qV}{kT}\right)^{-1}}{1 + \dfrac{S}{D_n}\left(\dfrac{2}{W}\dfrac{\varphi_o - qV}{kT}\right)^{-1}} - \exp(-\alpha W).$$ (7)

where S is the recombination velocity at the front surface, D_n is the electron diffusion coefficient related to the electron mobility μ_n through the Einstein relation: $qD_n/kT = \mu_n$.
For the *diffusion* component of the photoelectric quantum yield that takes into account surface recombination at the back surface of the CdTe layer, one can use the exact expression obtained for the p-layer in a p-n junction solar cell (Sze, 1981)

$$\eta_{dif} = \frac{\alpha L_n}{\alpha^2 L_n^2 - 1}\exp(-\alpha W) \times$$

$$\times \left\{ \alpha L_n - \frac{\dfrac{S_b L_n}{D_n}\left[\cosh\left(\dfrac{d-W}{L_n}\right) - \exp(-\alpha(d-W))\right] + \sinh\left(\dfrac{d-W}{L_n}\right) + \alpha L_n \exp(-\alpha(d-W))}{\dfrac{S_b L_n}{D_n}\sinh\left(\dfrac{d-W}{L_n}\right) + \cosh\left(\dfrac{d-W}{L_n}\right)} \right\},$$ (8)

where d is the thickness of the CdTe absorber layer, S_b is the recombination velocity at its back surface.
The *total* quantum yield of photoelectric conversion in the CdTe absorber layer is the sum of the two components: $\eta_{int} = \eta_{drift} + \eta_{dif}$.
Fig. 14 illustrates a comparison of the calculated curve $\eta_{ext}(\lambda)$ using Eqs. (5)-(8) with the measured spectrum (Kosyachenko et al., 2009). As seen, very good agreement between the calculated curve and the experimental points has been obtained.

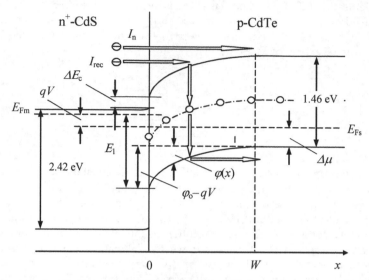

Fig. 13. The energy band diagram of CdS/CdTe thin-film heterojunction under forward bias. The electron transitions corresponding to the recombination current I_{rec} and over-barrier diffusion current I_n are shown.

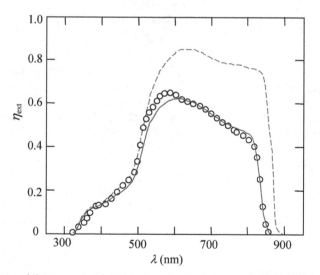

Fig. 14. Comparison of the measured (circles) and calculated (solid line) quantum efficiency spectrum η_{ext}. The dashed line shows the spectrum of 100 % *internal* efficiency.

The expressions for quantum efficiency spectra can be used to calculate the short-circuit current density J_{sc} using AM1.5 solar radiation Tables ISO 9845-1:1992 (Standard ISO, 1992). If Φ_i is the spectral radiation power density and $h\nu$ is the photon energy, the spectral density of the incident photon flux is $\Phi_i/h\nu_i$ and then

$$J_{sc} = q \sum_i \eta_{int}(\lambda) \frac{\Phi_i(\lambda)}{h\nu_i} \Delta\lambda_i , \qquad (9)$$

where $\Delta\lambda_i$ is the wavelength range between the neighboring values of λ_i (the photon energy $h\nu_i$) in the table and the summation is over the spectral range 300 nm $\le \lambda \le \lambda_g = hc/E_g$.

The calculation results of the *drift* component of short-circuit current density J_{drift} using Eqs. (7) and (9) lead to important practical conclusions (Kosyachenko et al., 2008).

If $S = 0$, the short-circuit current gradually increases with widening W and approaches a maximum value $J_{drift} = 28.7$ mA/cm² at $W > 10$ μm. Surface recombination decreases J_{drift} only in the case if the electric field in the space-charge region is not strong enough, i.e. when the uncompensated acceptor concentration $N_a - N_d$ is low. As $N_a - N_d$ increases and consequently the electric field strength becomes stronger, the influence of surface recombination becomes weaker, and at $N_a - N_d \ge 10^{16}$ cm⁻³ the effect of surface recombination is virtually eliminated. However, in this case, J_{drift} decreases with increasing $N_a - N_d$ because a significant portion of radiation is absorbed outside the space-charge region. Thus, the dependence of drift component of the short-circuit current on the uncompensated acceptor concentration $N_a - N_d$ is represented by a curve with a maximum at $N_a - N_d \approx 10^{15}$ cm⁻³ ($W \approx 1$ μm).

The *diffusion* component of short-circuit current density J_{dif} is determined by the thickness of the absorber layer d, the electron lifetime τ_n and the recombination velocity at the back surface of the CdTe layer S_b. If, for example, $\tau_n = 10^{-6}$ s and $S_b = 0$, then the total charge collection in the neutral part is observed at $d = 15-20$ μm and to reach the total charge collection in the case $S_b = 10^7$ cm/s, the CdTe thickness should be 50 μm or larger (Kosyachenko et al., 2008). In this regard the question arises why for total charge collection the thickness of the CdTe absorber layer d should amount to several tens of micrometers. The matter is that, as already noted, the value of d is commonly considered to be in excess of the effective penetration depth of the radiation into the CdTe absorber layer in the intrinsic absorption region of the semiconductor, i.e. in excess of $d = 10^{-4}$ cm $= 1$ μm. With this reasoning, the absorber layer thickness is usually chosen at a few microns. However, one does not take into account that the carriers, arisen outside the space-charge region, *diffuse* into the neutral part of the CdTe layer penetrating deeper into the material. Having reached the back surface of the CdTe layer, carriers recombine and do not contribute to the photocurrent. Considering the spatial distribution of photogenerated electrons in the neutral region shows that at $S_b = 7 \times 10^7$ cm/s, typical values of $\tau_n = 10^{-9}$ s and $N_a - N_d = 10^{16}$ cm⁻³ and at $d = 1-2$ μm, surface recombination "kills" most of electrons photogenerated in the neutral part of the CdTe layer (Kosyachenko et al., 2009).

Fig. 15 shows the calculation results of the *total* short-circuit current density J_{sc} (the sum of the drift and diffusion components) vs. $N_a - N_d$ for different electron lifetimes τ_n. Calculations have been carried out for the CdTe film thickness $d = 5$ μm which is often used in the fabrication of CdTe-based solar cells. As can be seen, at $\tau_n \ge 10^{-8}$ s the short-circuit current density is 26-27 mA/cm² when $N_a - N_d > 10^{16}$ cm⁻³ and for shorter electron lifetime, J_{sc} peaks at $N_a - N_d = (1-3) \times 10^{15}$ cm⁻³.

As $N_a - N_d$ is in excess of this concentration, the short-circuit current decreases since the drift component of the photocurrent reduces. In the range of $N_a - N_d < (1-3) \times 10^{15}$ cm⁻³, the short-circuit current density also decreases, but due to recombination at the front surface of the CdTe layer.

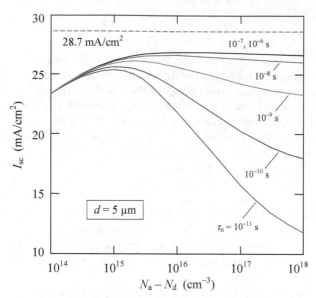

Fig. 15. *Total* short-circuit current density J_{sc} of a CdTe-based solar cell as a function of the uncompensated acceptor concentration N_a - N_d calculated at the absorber layer thickness d = 5 μm for different electron lifetime τ_n.

The *I-V* characteristic determined the open-circuit voltage and fill factor of CdS/CdTe solar cells is most commonly described by the semi-empirical formulae similar to Eq. (4), which consists the so-called "ideality" factor and is valid for some cases. Our measurements show, however, that such "generalization" of the formulae does not cover the observed variety of the CdS/CdTe solar cell *I-V* characteristics. The measured voltage dependences of the forward current are not always exponential and the saturation of the reverse current is *never* observed.

On the other hand, our measurements show that the *I-V* characteristics of CdS/CdTe heterostructures and their temperature variation are governed by the generation-recombination Sah-Noyce-Shockley theory (Sah at al., 1957). According to this theory, the dependence $I \sim \exp(qV/nkT)$ at $n \approx 2$ takes place only in the case, where the generation-recombination energy level is located near the middle of the band gap. If the level moves away from the mid-gap the coefficient n becomes close to 1 but only at low forward voltage. If the forward voltage elevates, the *I-V* characteristic modifies in the dependence where $n \approx 2$ and at higher voltages the dependence I on V becomes even weaker (Sah et al., 1957; Kosyachenko et al., 2004). Certainly, at higher forward currents, it is also necessary to take into account the voltage drop across the series resistance R_s of the bulk part of the CdTe layer by replacing the voltage V in the discussed expressions with $V - I \cdot R_s$.

The Sah-Noyce-Shockley theory supposes that the generation-recombination rate in the space-charge region is determined by expression

$$U(x,V) = \frac{n(x,V)p(x,V) - n_i^2}{\tau_{po}\left[n(x,V) + n_1\right] + \tau_{no}\left[p(x,V) + p_1\right]},$$ (10)

where $n(x,V)$ and $p(x,V)$ are the carrier concentrations in the conduction and valence bands, n_i is the intrinsic carrier concentration. The n_1 and p_1 values in Eq. (10) are determined by the energy spacing between the top of the valence band and the generation-recombination level E_t, i.e. $p_1 = N_v \exp(-E_t/kT)$ and $n_1 = N_c \exp[-(E_g - E_t)/kT]$, where $N_c = 2(m_n kT/2\pi\hbar^2)^{3/2}$ and $N_v = 2(m_p kT/2\pi\hbar^2)^{3/2}$ are the effective density of states in the conduction and valence bands, m_n and m_p are the effective masses of electrons and holes, and τ_{no} and τ_{po} are the lifetime of electrons and holes in the depletion region, respectively.

The recombination current under forward bias and the generation current under reverse bias are found by integration of $U(x, V)$ throughout the entire depletion layer:

$$J_{gr} = q \int_0^W U(x,V)dx . \tag{11}$$

In Eq. (10) the expressions for $n(x,V)$ and $p(x,V)$ in the depletion region have the forms:

$$p(x,V) = N_c \exp\left[-\frac{\Delta\mu + \varphi(x,V)}{kT}\right], \tag{12}$$

$$n(x,V) = N_v \exp\left[-\frac{E_g - \Delta\mu - \varphi(x,V) - qV}{kT}\right], \tag{13}$$

where $\Delta\mu$ is the energy spacing between the Fermi level and the top of the valence band in the bulk of the CdTe layer, $\varphi(x, V)$ is the potential energy given by Eq. (5).

Over-barrier (diffusion) carrier flow in the CdS/CdTe heterostructure is restricted by high barriers for both majority carriers (holes) and minority carriers (electrons) (Fig. 13). That is why, under low and moderate forward voltages, the dominant charge transport mechanism is caused by recombination in the space-charge region. However, as qV nears φ_o, the over-barrier currents due to much stronger dependence on V become comparable and even higher than the recombination current. Since in CdS/CdTe heterojunction the barrier for holes is considerably higher than that for electrons, the *electron* component dominates the over-barrier current, which can be written as (Sze, 1981):

$$J_n = q \frac{n_p L_n}{\tau_n}\left[\exp\left(\frac{qV}{kT}\right) - 1\right], \tag{14}$$

where $n_p = N_c \exp[-(E_g - \Delta\mu)/kT]$ is the concentration of electrons in the neutral part of the p-CdTe layer.

Thus, the dark current density $J_d(V)$ in CdS/CdTe heterostructure is the sum of the generation-recombination and diffusion components:

$$J_d(V) = J_{gr}(V) + J_n(V) . \tag{15}$$

The results of comparison between theory and experiment are demonstrated in Fig. 16 on the example of *I-V* characteristic, which reflects especially pronounced features of the transport mechanism in CdS/CdTe solar cell (Kosyachenko et al., 2010). As is seen, there is an extended portion of the curve ($0.1 < V < 0.8$ V), where the dependence $I \sim \exp(qV/AkT)$ holds for $n = 1.92$ (rather than 2!).

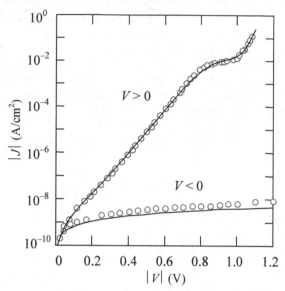

Fig. 16. Room-temperature *I-V* characteristic of thin-film CdS/CdTe heterostructure. The circles and solid lines show the experimental and calculated results, respectively.

At higher voltages, the deviation from the exponential dependence toward lower currents is observed. However, if the voltage elevates still further (> 1 V), a much steeper increase of forward current occurs. Analysis shows that all these features are explained in the frame of mechanism involving the generation-recombination in the space-charge region in a wide range of moderate voltages completed by the over-barrier diffusion current at higher voltages.

One can see in Fig. 16 that the *I-V* characteristic calculated in accordance with the above theory are in good agreement with experiment both for the forward and reverse connections of the solar cell. Note that the reverse current increases continuously with voltage rather than saturates, as requires the commonly used semi-empirical formula.

Knowing the dark *I-V* characteristic, one can find the *I-V* characteristic under illumination as

$$J(V) = J_d(V) - J_{ph} \tag{16}$$

and determine the open-circuit voltage and fill factor. In Eq.(16) $J_d(V)$ and J_{ph} are the dark current and photocurrent densities, respectively. Of course, it must be specified a definite value of the density of short circuit current J_{sc}. Keeping in view the determination of conditions to maximize the photovoltaic efficiency, we use for this the data shown in Fig. 15, i.e. set $J_{sc} \approx 26$ mA/cm². This is the case for $N_a - N_d = 10^{15}$-10^{16} cm⁻³ and a film thickness $d = 5$ μm, which is often used in the fabrication of CdTe-based solar cells.

Fig. 17 shows the open-circuit voltage V_{oc} and the efficiency η of CdS/CdTe heterostructure as a function of effective carrier lifetime τ calculated for various resistivities of the p-CdTe layer ρ.

As seen in Fig. 17(a), the open-circuit voltage V_{oc} considerably increases with lowering ρ and increasing τ (as ρ varies, $\Delta\mu$ also varies affecting the value of the recombination current, and especially the over-barrier current).

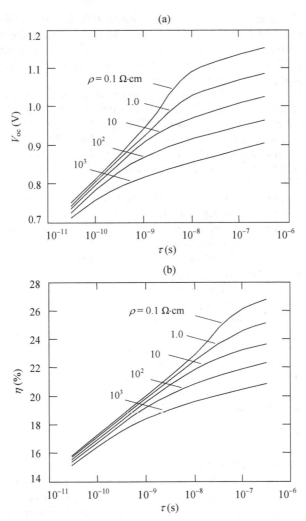

Fig. 17. Dependences of the open-circuit voltage V_{oc} (a) and efficiency η (b) of CdS/CdTe heterojunction on the carrier lifetime τ calculated by Eq. (16) using Eqs. (10)-(15) for various resistivities ρ of the CdTe layer.

In the most commonly encountered case, as $\tau = 10^{-10}$-10^{-9} s, the values of $V_{oc} = 0.8$-0.85 V (0.75-0.8 V for commercial devices) are far from the maximum possible values of 1.15-1.2 V, which are reached on the curve for $\rho = 0.1$ $\Omega\cdot$cm and $\tau > 10^{-8}$ s.

As seen in Fig. 17(b), the dependence of the efficiency $\eta = P_{out}/P_{irr}$ on τ remarkably increases from 15-16% to 21-27.5% when τ and ρ changes within the indicated limits (P_{irr} is the AM 1.5 solar radiation power 100 mW/cm²). For $\tau = 10^{-10}$-10^{-9} s, the efficiency lies near 17-19% and the enhancement of η by lowering ρ of the CdTe layer is 0.5-1.5%. Thus, assuming $\tau = 10^{-10}$-10^{-9} s, the calculated results turn out to be quite close to the experimental efficiencies of the best thin-film CdS/CdTe solar cells (16-17%).

The enhancement of η from 16-17% to 27-28% is possible if the carrier lifetime increases to $\tau \geq 10^{-6}$ s and the resistivity of CdTe reduces to $\rho \approx 0.1$ $\Omega \cdot$cm. This also requires an increase in the short-circuit current density. As follows from the foregoing, the latter is possible for the thickness of the CdTe absorber layer of 20-30 μm and even more. Evidently, this is not justified for large-scale production of solar modules.

In the early years of 21 century, the technology and manufacturing of solar modules based on CdTe, which could compete with silicon counterparts was developed. With mass production, the efficiency of CdTe modules is 10-11% with the prospect of an increase in a few percents in the coming years (Multi Year Program Plan, 2008). The cost of modules over the past five years has decreased three times and crossed the threshold $1.0 per Wp, that is much less than wafer or ribbon based modules on silicon. In 2012-2015, the cost of CdTe-based solar modules is expected to be below $ 0.7 per Wp.

It should be noted that the growth rates of CdTe module production over the last decade are the highest in the entire solar energy sector. Over the past 5 years, their annual capacity increased more than an order of magnitude, greatly surpassing the capacity of the counterparts based on a-Si and in a few times – based on CIS (CIGS). In Germany, Spain, USA and other countries, CdTe solar photovoltaic power plants with a capacity of several megawatts up to several tens of megawatts have been built. Annual production of solar modules based on CdTe by only one company First Solar, Inc. in 2009-2010 exceeded 1.2 GW). This company is the largest manufacturer of solar modules in the world, which far exceeded the capacities of perennial leaders in the manufacture of solar modules and continues to increase production, despite the economic and financial crisis. Other well known companies such as AVA Solar and Prime Star Solar (USA), Calyxo GmbH and Antec Solar Energy AG (Germany), Arendi SRL (Italy) are also involved in the production of CdTe solar modules. In May 2010 the General Electric company announced plans to introduce production of CdTe thin-film solar modules based on technology developed at the National Renewable Energy Laboratory and PrimeStar Solar. These facts remove any doubt on the prospects of solar energy based on CdTe.

One of the arguments advanced against the use of CdTe in solar energy is based on the fact that natural resources of Cd and Te are limited.

Indeed, Cd and Te are rare and scattered elements; their content in the earth's crust is $\sim 10^{-5}$% and $\sim 10^{-7}$-10^{-6}%, respectively. Currently, there are no commercial deposits of Cd and Te in the world; Cd and Te are extracted as byproducts in the production of mainly zinc and copper, respectively. The limiting raw factor for development of solar energy through the production of CdTe is Te. For the world needs, cadmium is annually produced just 150-200 tons. According to the National Renewable Energy Laboratory, the U.S. Department of Energy and other agencies, annual production of Te as a byproduct of copper production can be increased to ~ 1.5 tons. For the module production with capacity of 1 GW, approximately 70 tons of Te are needed at present 10-11% efficiency of modules. Using each year, for example, 1 thousand tons one can make solar modules with power ~ 15 GW. Thus, through Te only as a byproduct in the production of Cu, accelerated development of solar energy based on CdTe can last for several decades. Other currently unused stocks of tellurium, particularly in South America, China, Mexico and other places of the globe are also known. With good reason Te was not the focus of geological exploration, however, studies in recent years show that, for example, underwater crusts throughout the ocean basins is extremely rich in Te, whose content of Te $\sim 10^9$ times higher compared with ocean water and $\sim 10^4$ times higher than in the Earth's crust (Hein et al., 2003). These stocks of tellurium in a relatively small depth of ocean (e.g. ~ 400 m) can easily meet the needs of the whole world's energy. It should also be noted that the additional costs of Cd and Te will not arise after 25-35 years, when CdTe solar panels expend their resources. The technologies for *recycling* the worked-out products, which allows majority of the components (~ 90%) to use in the production of new solar modules, have been already developed.

Another objection to the proliferation of CdTe solar cells, which opponents argue, is that the Cd, Te and their compounds are extremely harmful to humans.

Indeed, Cd and Te are toxic heavy metals; Cd is even cancer-causing element. However, the research of many independent experts of the National Renewable Energy Laboratory and Brookhaven National Laboratory show that CdTe compound is chemically stable, biologically inert and does not constitute a threat to human health and the environment both in terms of production and exploitation of solar modules (Bonnet, 2000; Fthenakis, 2008). Cd emissions to the atmosphere is possible only if the temperature exceeds ~ 1050°C in case of fire. However, CdTe in solar module is between two glass plates in a sealed condition. With this design, glass will melt at temperatures much lower than 1050°C, CdTe will turn in the molten mass that does not allow the allocation of Cd and Te in the atmosphere. It has been shown that the release of cadmium to the atmosphere is lower with CdTe-based solar cells than with silicon photovoltaics. Despite much discussion of the toxicity of CdTe-based solar cells, this is technology that is reliably delivered on a large scale.

4. Conclusions

Analysis of photovoltaics development leads to the negative conclusion that the desired rate of increase in the capacity of solar energy based on single-crystalline, polycrystalline and amorphous silicon can not be provided. Despite a long history, the share of PV currently amounts to a small fraction of the overall balance of the world power sector, and even according to the most optimistic forecasts, will not dominate in 2050. Resources of hydroelectric and wind energy are limited, the expansion of nuclear power is highly problematic from a security standpoint. This means that a significant fraction of the energy will be generated by natural gas, oil, coal, oil shale, biomass, which can lead to irreversible changes in climate on Earth. The main reason for the slow development of the photovoltaics based on wafer or ribbon silicon (as its main direction) is the high consumption of materials, energy and labor, and hence too low productivity and high cost of production. This is determined by the *fundamental* factor because the single-crystalline and polycrystalline silicon are indirect-gap semiconductors. The technology of solar modules based on direct-gap amorphous silicon is quite complicated, and their stabilized efficiency is too low for use in large-scale energy. In this regard, there is an urgent need to involve other areas of photovoltaics in energy production. Thin-film technologies using direct-gap semiconductors such as CIGS and CdTe hold the promise of significantly accelerating the development of photovoltaics. Intensive research and development of thin-film technologies based on other materials, for example, organic and die-sensitizes solar cells is also being conducted. The main advantages of thin-film technology are less material consumption, lower requirements to the parameters of the materials, ease of engineering methods of manufacture, and the possibility of full automation. All of this provides better throughput of manufacturing and lower production costs, i.e. just what is lacking in wafer or ribbon based silicon photovoltaics. CdTe and CIGS based modules have proved their viability. Solar power stations based on these materials with a capacity from a few megawatts to a few tens of megawatts have already been built; several agreements for the construction of such plants with a capacity higher by one or even two orders of magnitude have been concluded. A growing number of companies are involved in the production of CdTe and CIGS based modules. Broad front of research on the possibility of increasing the efficiency of the modules, which in mass production is much lower than the theoretical predictions, are being conducted. The aforesaid, of course, does not preclude participation in the production of electrical energy of photovoltaics based on single-crystalline, polycrystalline, ribbon and amorphous silicon with different designs of solar cell structures. A large number of companies are involved in the production of the silicon modules, which are continually evolving, making a potential contribution to the energy, but they cannot solve the problem globally for the foreseeable future.

5. Acknowledgements

I thank X. Mathew, Centro de Investigacion en Energia-UNAM, Mexico, for the CdS/CdTe thin-film heterostructures, V.M. Sklyarchuk for sample preparation to study, E.V. Grushko for measurements and all participants of the investigation for helpful discussion. The study was supported by the State Foundation for Fundamental Investigations of Ukraine within the Agreements $\Phi14/259$-2007 and $\Phi40.7/014$.

6. References

Adirovich, E.I., Yuabov, Yu.M., Yagudaev, G.R. (1969). Photoelectric phenomena in film diodes with heterojunction. *Sov. Phys. Semicond.* 3, 61-65

Alonso, M.I., Garriga, M., Durante Rincón, C.A., Hernández, E. and León, M. (2002). Optical functions of chalcopyrite $CuGa_xIn_{1-x}Se_2$ alloys. *Applied Physics A: Materials Science & Processing*, 74, 659-664.

Basore, P.A. (2006). CSG-2: Expanding the production of a new polycrystalline silicon PV technology. *Proc. of the 21st European Photovoltaic Solar Energy Conference.*

Birkmire, R.W. & Eser, E. (1997). Polycrystalline thin film solar cells: Present status and future potential, *Annu. Rev. Mater. Sc.* 27, 625.

Birkmire, R.W. (2008). Pathways to improved performance and processing of CdTe and $CuInSe_2$ based modules. *Conference Record of 33rd IEEE Photovoltaic Specialists Conference.* pp. 47-53.

Bonnet, D. (2001). Cadmium telluride solar cells. In: *Clean Electricity from Photovoltaics.* Ed. By M.D. Archer and R. Hill. Imperial College Press, New York. pp. 245-275.

Bonnet, D., Oelting, S., Harr, M., Will, S. (2002). Start-up and operation of an integrated 10MWp thin film PV module factory. *Proceedings of the 29th IEEE Photovoltaic Specialists Conference*, pp. 563–566.

Bonnet, D. (2003). Cadmium telluride thin-film PV modules. in: *Practical Handbook of Photovoltaics: Fundamentals and Applications*, edited by T. Markvart and L. Castaner, (Elseiver, New York,). p. 333-366.

Britt, J. & Ferekides, C. (1993). Thin Film CdS/CdTe Solar Cell with 15.8% efficiency. *Appl. Phys. Lett.* 62, 2851-2852.

Carlson, D.E. & Wronski, C.R. (1976). Amorphous silicon solar cell. *Appl. Phys. Lett.* 28, 671-672

Chamberlain, G.A. (1983). Organic solar cells: a review. *Solar Cells,* 8, 47-83.

Chiba, Y.; Islam, A.; Watanabe, Y.; Koyama, R.; Koide, N.; & Han, L. (2006). Dye-Sensitized Solar Cells with Conversion Efficiency of 11.1% *Jpn. J. Appl. Phys.* 45, L638-L640.

Chu, T.L, Chu, S., Lin, S., Yue, J. (1984). Large Grain Copper Indium Didiselenide Films. *J. Electrochemical Soc.* 131, 2182-2185.

Contreras, M.A., Egaas, B., Ramanathan, K., Hiltner, J., Swartzlander, A., Hasoon, F., Noufi, R. Progress towards 20% efficiency in $Cu(In,Ga)Se_2$ polycrystalline thin-film solar cells. (1999). *Prog. Photovolt: Res. Appl.* 7, 311-316.

Deckman, H.W., Wronski, C.R., Witzke, H. and Yablonovitch, E. (1983). Optically enhanced amorphous silicon solar cells. *Appl. Phys. Lett.* 42, 968-970.

Deng, S. & Schiff, E.A. (2003). Amorphous Silicon-based Solar Cells. In: *Handbook of Photovoltaic Science and Engineering*, Second Edition. Edited by A. Luque and S. Hegedus. John Wiley & Sons, Ltd., pp. 505-565.

EPIA Global market outlook for photovoltaics until 2015. (2011). European Photovoltaic Industry Association. www.epia.org.

EUR 24344 PV Status Report 2010. Institute for Energy. Joint Research Centre. European Commission. (2010). http://www.jrc.ec.europa.eu.

Fahrenbruch, A. & Bube, R., (1983). *Fundamentals of Solar Cells*, 231-234, Academic Press, New York.

Faraji, M., Sunil Gokhale, Choudhari, S. M., Takwale, M. G., and S. V. Ghaisas. (1992). High mobility hydrogenated and oxygenated microcrystalline silicon as a photosensitive material in photovoltaic applications. *Appl. Phys. Lett.* 60, 3289-3291.

Ferrazza, F. (2003). Silicon: manufacture and properties. In: *Practical Handbook of Photovoltaics: Fundamentals and Applications*, edited by T. Markvart and L. Castaner (Elseiver, New York), pp. 137-154.

Fritsche, J., Kraft, D., Thissen, A., Mayer, Th., Klein & A., Jaegermann W. (2001). Interface engineering of chalcogenide semiconductors in thin film solar cells: CdTe as an example, *Mat. Res. Soc. Symp. Proc.* , 668, 601-611.

Fthenakis, V.M., Kim, H.C., and Alsema, E. (2008). Emissions from Photovoltaic Life Cycles. *Environ. Sci. Technol.* 42, 2168-2174.

Goetzberger, A. , Hebling, C. & Schock, H.-W. (2003). Photovoltaic materials, history, status and outlook. *Materials Science and Engineering* R40, 1-46.

Cohen-Solal, G., Lincot, D., Barbe, M. (1982). High efficiency shallow p(+)nn(+) cadmium telluride solar cells. in: *Photovoltaic Solar Energy Conference*; Proceedings of the 4th International Conference, 621-626.

Cooke, M. (2008). CdTe PV progresses to mass production. *Semiconductors Today*, 3, 74-77.

Grätzel, M. (2003). Dye-sensitized solar cells. *J. Photochem. Photobiol. C: Photochem. Rev.* 4, 145-153.

Gratzel, M. (2004). Conversion of sunlight to electric power by nanocrystalline dye-sensitized solar cells. *J. Photoch. & Photobiology* A164, Issue: 1-3, 3-14.

Gray, J.L., Schwartz, R.J., Lee, Y.J. (1990). Development of a computer model for polycrystalline thin-film $CuInSe_2$ and CdTe solar cells. *Annual report 1 January – 31 December 1990*. National Renewable Energy Laboratory/TP-413-4835, 1-36.

Green, M.A., Emery, K., Hishikawa, K.Y. & Warta, W. (2011). Solar Cell Efficiency Tables (version 37). *Progress in Photovoltaics: Research and Applications* 19, pp. 84-92. ISSN 1099-159X.

Guter, W., Schöne, J., Philipps, S., Steiner, M., Siefer, G., Wekkeli, A., Welser, E, Oliva, E, Bett, A., and Dimroth F.(2009). Current-matched triple-junction solar cell reaching 1.1% conversion efficiency under concentrated sunlight, *Appl. Phys. Lett.* 94, 023504 (8 pages).

Han, S.-H., Hasoon, F.S., Hermann, A.M., Levi, D.H. (2007). Spectroscopic evidence for a surface layer in CuInSe$_2$:Cu deficiency. *Appl. Phys. Lett.* 91, 021904 (5 pages).

Haugeneder, A.; Neges, M.; Kallinger, C.; Spirkl, W.; Lemmer, U.; Feldmann, J.; Scherf, U.; Harth, E.; Gügel, A.; Müllen, K. (1999). Exciton diffusion and dissociation in conjugated polymer/fullerene blends and heterostructures. *Phys. Rev. B.*, 59, 15346-15351.

Hegedus, S. & Luque, A. (2011). Achievements and Challenges of Solar Electricity from Photovoltaics. In: *Handbook of Photovoltaic Science and Engineering*, Second Edition. Edited by A. Luque and S. Hegedus. John Wiley & Sons, Ltd., pp. 2-38.

Hein, J.R., Koschinsky, A., and Halliday, A.N. (2003). Global occurrence of tellurium-rich ferromanganese crusts and a model for the enrichment of tellurium. *Geochimica et Cosmochimica Acta*, 67, 1117-1127.

http://refractiveindex.info/?group=CRYSTALS&material=a-Si.

http://www1.eere.energy.gov/solar/pdfs/solar_program_mypp_2008-2012.pdf. *Multi Year Program Plan 2008-2012*. U.S. Department of Energy.

Jager-Waldau, A. (2010). Research, Solar Cell Production and Market Implementation of Photovoltaics. *Luxembourg: Office for Official Publications of the European Union.* 120 pp.

Ji, J. J. & Shi, Z. (2002). Texturing of glass by SiO$_2$ film. *US patent 6, 420, 647.*

Jordan, J. (1993). *International Patent Application.* WO93/14524.

Kazmerski, L.L., White, F.R., and Morgan, G.K. (1976). Thin-film CuInSe$_2$/CdS heterojunction solar cells. *Appl. Phys. Lett.* 29, 268-270.

Keppner, H., Meier, J. Torres, P., Fischer, D., and Shah, A. (1999). Microcrystalline silicon and micromorph tandem solar cells. *Applied Physics A: Materials Science & Processing.* 69, 169-177.

King, R. (2008). Multijunction cells: record breakers, *Nature Photonics*, 2, 284– 285.

Kosyachenko, L.A., Maslyanchuk, O.L., Motushchuk, V.V., Sklyarchuk, V.M. (2004). Charge transport generation-recombination mechanism in Au/n-CdZnTe diodes. *Solar Energy Materials and Solar Cells.* 82, 65-73.

Kosyachenko, L.A., Grushko, E.V., Savchuk, A.I. (2008). Dependence of charge collection in thin-film CdTe solar cells on the absorber layer parameters. *Semicond. Sci. Technol.* 23, 025011 (7pp).

Kosyachenko, L., Lashkarev, G., Grushko, E., Ievtushenko, A., Sklyarchuk, V., Mathew X., and Paulson, P.D. (2009). Spectral Distribution of Photoelectric Efficiency of Thin-Film CdS/CdTe Heterostructure. *Acta Physica Polonica.* A 116 862-864.

Kosyachenko, L.A., Savchuk, A.I., Grushko, E.V. (2009). Dependence of the efficiency of a CdS/CdTe solar cell on the absorbing layer's thickness. *Semiconductors*, 43, 1023-1027.

Kosyachenko, L.A., Grushko, E.V. (2010). Open_Circuit Voltage, Fill Factor, and Efficiency of a CdS/CdTe Solar Cell. *Semiconductors*, 44, 1375–1382.

Kuwano, Y., Ohniishi, Nishiwaki, H., Tsuda, S., Fukatsu, T., Enomoto, K., Nakashima, Y., and Tarui, H. (1982). Multi-Gap Amorphous Si Solar Cells Prepared by the Consecutive, Separated Reaction Chamber Method. *Conference Record of the 16th IEEE Photovoltaic Specialists Conference*, (San Diego, CA, 27-30.9.1982), pp. 1338-1343.

Lavagna, M., Pique, J.P. & Marfaing, Y. (1977). Theoretical analysis of the quantum photoelectric yield in Schottky diodes, *Solid State Electronics*, 20, 235-240.

Loferski, J.J. (1956). Theoretical Considerations Governing the Choice of the Optimum Semiconductor for Photovoltaic Solar Energy Conversion. *J. Appl. Phys.* 27, 777-784.

Mauk, M., Sims, P., Rand, J., and Barnett. (2003). A. Thin silicon solar cells In: *Practical Handbook of Photovoltaics: Fundamentals and Applications*, edited by T. Markvart and L. Castaner (Elseiver, New York), p. 185-226.

Meier, J., Flückiger, R., Keppner, H., and A. Shah. (1994). Complete microcrystalline p-i-n solar cell—Crystalline or amorphous cell behavior? *Appl. Phys. Lett.* 65, 860-862.

Meier, J., Dubail, S., Fischer, D., Anna Selvan, J.A., Pellaton Vaucher, N., Platz, Hof, C., Flückiger, R., Kroll, U., Wyrsch, N., Torres, P., Keppner, H., Shah, A., Ufert, K.-D. (1995). The 'Micromorph' Solar Cells: a New Way to High Efficiency Thin Film Silicon Solar Cells. *Proceedings of the 13th EC Photovoltaic Solar Energy Conference*, Nice, pp. 1445-1450.

Meier, J., Dubail, S., Cuperus, J., Kroll, U., Platz, R., Torres, P., Anna Selvan, J.A., Pernet, P., Beck, N., Pellaton Vaucher, N., Hof, Ch., Fischer, D., Keppner, H., Shah, A. (1998). Recent progress in micromorph solar cells. *J. Non-Crystalline Solids*. 227-230, 1250-1256.

Meyers, P.V. & Albright, S.P. (2000). Technical and Economic opportunities for CdTe PV at the turn of Millennium. *Prog. Photovolt.: Res. Appl.* 8, 161-169.

Mickelsen, RA. & Chen, W.S. (1981). Development of a 9_4% efficient thin-film $CuInSe_2$/CdS solar cell. *Proc.15th IEEE Photovoltaic Specialists Conference*, New York,; 800-804.

Muller, R.S. & Zuleeg, R. (1964). Vapor-Deposited, Thin-Film Heterojunction Diodes. *J. Appl. Phys.* 35, 1550-1558.

Multi Year Program Plan 2008-2012. U.S. Department of Energy. http://www1.eere.energy.gov/ solar/pdfs/solar_program_mypp_2008-2012.pdf

Minilya-Arroyo, J., Marfaing, Y., Cohen-Solal, G., Triboulet, R. (1979). Electric and photovoltaic properties of CdTe p-n homojunctions. *Sol. Energy Mater.* 1, 171-180.

Nakada, T. & Mise, T. (2001). High-efficiency superstrate type cigs thin film solar cells with graded bandgap absorber layers. *Proceedings of the 17th European Photovoltaic Solar Energy Conference*, Munich. 1027-1030.

Nishitani, M., Negami, T., Kohara, N., and Wada, T. (1997). Analysis of transient photocurrents in Cu(In,Ga)Se2 thin film solar cells. *J. Appl. Phys.* 82, 3572-3575.

Neumann, H., Tomlinson, R.D. (1990). Relation between electrical properties and composition in $CuInSe_2$ single crystals. *Sol. Cells.* 28, 301-313.

Ohnesorge, B., Weigand, R., Bacher, G., Forchel, A., Riedl, W., and Karg, F. H. (1998). Minority-carrier lifetime and efficiency of Cu(In,Ga)Se2 solar cells. *Appl. Phys Lett.* 73, 1224-1227.

O'Regan, B. & Grätzel, M. (1991). A Low-Cost, High-Efficientcy Solar-Cell Based on Dye-Sensitized Colloidal TiO_2 Films, *Nature*, 353, No. 6346, 737-740.

Paulson, P.D. , Birkmire, R.W. and Shafarman, W.N. (2003). Optical Characterization of CuIn1-xGaxSe2 Alloy Thin Films by Spectroscopic Ellipsometry. *J. Appl. Phys.*, 94, 879-888.

Repins, I., Contreras, M.A., Egaas, B., DeHart, C., Scharf, J., Perkins, C.L., To, B., and Noufi. R. (2008). 19 9%-efficient ZnO/CdS/CuInGaSe2 solar cell with 81.2% fill factor. *Progress in Photovoltaics: Research and Applications*. 16, 235-239.

Romeo, A., M. Terheggen, Abou-Ras, D., Bőtzner, D.L., Haug, F.-J., Kőlin, M., Rudmann, D., and Tiwari, A.N. (2004). Development of Thin-film Cu(In,Ga)Se2 and CdTe Solar Cells. *Prog. Photovolt: Res. Appl.*;12, 93–111.

Spear, W.E. & Le Comber, P.G. (1975). Substitutional doping of amorphous silicon. *Solid State Commun.* 17, 1193-1196.

Schmid, D., Ruckh, M., Grunvald, M., Schock. H. (1993). Chalcopyrite/defect chalcopyrite heterojunctions on the basis $CuInSe_2$. *J. Appl. Phys.* 73, 2902-2909.

Schroeder, D.J., Rockett, A.A. (1997). Electronic effects of sodium in epitaxial $CuIn_{1-x}Ga_xSe_2$. *J. Appl. Phys.* 82, 4982-4985.

Shafarman, W.N. & Stolt, L. (2003). Cu(InGa)Se2 Solar Cells. In: *Handbook of Photovoltaic Science and Engineering*, Second Edition. Edited by A. Luque and S. Hegedus. John Wiley & Sons, Ltd., pp. 567-616.

Sah C., Noyce R. & Shockley W. (1957). Carrier generalization recombination in p-n junctions and p-n junction characteristics, *Proc. IRE*. 45, 1228-1242.

Shah, J., Meier, A., Buechel, A., Kroll, U., Steinhauser, J., Meillaud, F., Schade, H. (2006). Towards very low-cost mass production of thin-film silicon photovoltaic (PV) solar modules on glass. *Thin Solid Film*. 502, 292-299.

Shay, J.L., Wagner, S., and Kasper, H.M. (1975). Efficient $CuInSe_2/CdS$ solar cells. *Appl. Phys. Lett.* 27, 89-90.

Staebler, D.L. & Wronski, C.R. (1977). Reversible conductivity changes in discharge-produced amorphous Si. *Appl. Phys. Lett.* 31, 292-294.

Standard of International Organization for Standardization ISO 9845-1:1992. Reference solar spectral irradiance at the ground at different receiving conditions.

Sze, S. (1981). *Physics of Semiconductor Devices*, 2nd ed. (Wiley, New York).

Szlufcik, J., Sivoththaman, S., Nijs, J.F., Mertens, R.P., and Overstraeten, R.V. (2003). Low cost industrial manufacture of crystalline silicon solar cells. In: *Practical Handbook of Photovoltaics: Fundamentals and Applications*, edited by T. Markvart and L. Castaner (Elseiver, New York), pp. 155-184.

Tang, C.W. (1986). 2-Layer Organic Photovoltaic Cell. *Appl. Phys. Lett.*, 48, 183-185.

Tiwari, A.N., Krejci, M., Haug, F.-J., Zogg, H. (1999). 12_8% Efficiency Cu(In, Ga)Se$_2$ solar cell on a flexible polymer sheet. *Progress in Photovoltaics: Research and Applications*. 7, 393-397.

Tuttle, J.R., Szalaj, A., Keane, J.A. (2000). 15_2% AMO/1433 W/kg thin-film Cu(In,Ga)Se$_2$ solar cell for space applications. *Proceedings of the 28th IEEE Photovoltaic Specialists Conference*, pp. 1042-1045.

Von Roedern, B., Kroposki, B., Strand, T. and Mrig, L. (1995). *Proccedings of the 13th European Photovoltaic Solar Energy Conference*. p. 1672-1676.

Von Roedern, B. & del Cueto, J.A. (2000). Model for Staebler-Wronski Degradation Deduced from Long-Term, Controlled Light-Soaking Experiments. *Materials Research Society's Spring Meeting*. San Francisco, California, April 24-28, 2000. 5 pages.

Von Roedern, B. (2006). Thin-film PV module review: Changing contribution of PV module technologies for meeting volume and product needs. *Renewable Energy Focus*. 7, 34-39.

Wagner, S., Shay, J. L., Migliorato, P., and Kasper, H.M. (1974). $CuInSe_2/CdS$ heterojunction photovoltaic detectors. *Appl. Phys. Lett.* 25, 434-435.

Widenborg, P.I. & Aberle, A.G. (2007). Polycrystalline Silicon Thin-Film Solar Cells on AIT-Textured Glass Superstrates. *Advances in OptoElectronics*. 24584 (7 pages).

Wu, X., Kane, J.C., Dhere, R.G., DeHart, C., Albin, D.S., Duda, A., Gessert. T.A., Asher, S., Levi. D.H., Sheldon, P. (2002). 16.5% Efficiency CdS/CdTe polycristalline thin-film solar cells. *Proceedings of the 17th European Photovoltaic Solar Energy Conference and Exhibition*, Munich, 995-1000.

www.firstsolar.com/recycling.

Yang, J, Banerjee, A, Guha, S. (1997). Triple-junction amorphous silicon alloy solar cell with 14.6% initial and 13.0% stable conversion efficiencies. *Appl. Phys. Lett.* 70, 2975-2977.

2

Low Cost Solar Cells Based on Cuprous Oxide

Verka Georgieva, Atanas Tanusevski[1] and Marina Georgieva
Faculty of Electrical Engineering and Information Technology,
[1]Institute of Physics, Faculty of Natural Sciences and Mathematics,
The "St. Cyril & Methodius" University, Skopje,
R. of Macedonia

1. Introduction

The worldwide quest for clean and renewable energy sources has encouraged large research activities and developments in the field of solar cells. In recent years, considerable attention has been devoted to the development of low cost energy converting devices. One of the most interesting products of photoelectric researches is the semiconductor cuprous oxide cell. As a solar cell material, cuprous oxide -Cu_2O, has the advantages of low cost and great availability. The potential for Cu_2O using in semiconducting devices has been recognized since, at least, 1920. Interest in Cu_2O revived during the mid seventies in the photovoltaic community (Olsen et al.,1982). Several primary characteristics of Cu_2O make it potential material for use in thin film solar cells: its non-toxic nature, a theoretical solar efficiency of about 9-11%, an abundance of copper and the simple and inexpensive process for semiconductor layer formation. Therefore, it is one of the most inexpensive and available semiconductor materials for solar cells. In addition to everything else, cuprous oxide has a band gap of 2.0 eV which is within the acceptable range for solar energy conversion, because all semiconductors with band gap between 1 eV and 2 eV are favorable material for photovoltaic cells (Rai, 1988).

A variety of techniques exist for preparing Cu_2O films on copper or other conducting substrates such as thermal, anodic and chemical oxidation and reactive sputtering. Particularly attractive, however, is the electrodeposition method because of its economy and simplicity for deposition either on metal substrates or on transparent conducting glass slides coated with highly conducting semiconductors, such as indium tin oxide (ITO), SnO_2, In_2O_3 etc. This offers the possibility of making back wall or front wall cells as well. We have to note that electrochemical preparation of cuprous oxide (Cu_2O) thin films has reached considerable attention during the last years.

Electrodeposition method of Cu_2O was first developed by Stareck (Stareck, 1937). It has been described by Rakhshani (Jayanetti & Dharmadasa, 1996, Mukhopadhyay et al.,1992, Rakhshani et al.1987, Rakhshani et al., 1996). In this work, a method of simple processes of electrolysis has been applied.

Electrochemical deposition technique is an simple, versatile and convenient method for producing large area devices. Low temperature growth and the possibility to control film thickness, morphology and composition by readily adjusting the electrical parameters, as well as the composition of the electrolytic solution, make it more attractive. At present, electrodeposition of binary semiconductors, especially thin films of the family of wide - bend gap II-IV semiconductors (as is ZnO), from aqueous solutions is employed in the preparation of solar cells. A photovoltaic device composed of a p-type semiconducting cuprous (I) oxide (Cu_2O) and n-type zinc oxide (ZnO) has attracted increasing attention as a future thin film solar cell, due to a theoretical conversion efficiency of around 18% and an absorption coefficient higher than that of a Si single crystal (Izaki et al. 2007)

Therefore, thin films of cuprous oxide (Cu_2O) have been made using electrochemical deposition technique. Cuprous oxide was electrodeposited on copper substrates and onto conducting glass coated with tin oxide (SnO_2), indium tin oxide (ITO) and zinc oxide (ZnO). Optimal conditions for high quality of the films were requested and determined. The qualitative structure of electrodeposited thin films was studied by x-ray diffraction (XRD) analysis. Their surface morphology was analyzed with scanning electronic microscope (SEM). The optical band gap values Eg were determined. To complete the systems Cu/Cu_2O, SnO_2/Cu_2O, ITO/Cu_2O and ZnO/Cu_2O as solar cells an electrode of graphite or silver paste was painted on the rear of the Cu_2O. Also a thin layer of nickel was vacuum evaporated on the oxide layer. The parameters of the solar cells, such the open circuit voltage (V_{oc}), the short circuit current (I_{sc}), the fill factor (FF), the diode quality factor (n), serial (Rs) and shunt resistant (Rsh) and efficiency (η) were determined. The barrier height (Vb) was determined from capacity-voltage characteristics.

Generally is accepted that the efficiency of the cells cannot be much improved. (Minami et al.,2004). But we successed to improve the stability of the cells, using thin layer of ZnO, making heterojunctions Cu_2O based cells.

2. Structural, morphological and optical properties of electrodeposited films of cuprous oxide

2.1 Experimental
2.1.1 Preparation of the films

A very simple apparatus was used for electrodeposition. It is consisted of a thermostat, a glass with solution, two electrodes (cathode and anode) and a standard electrical circuit for electrolysis. The deposition solution contained 64 g/l anhydrous cupric sulphate ($CuSO_4$), 200 ml/l lactic acid ($C_3H_6O_3$) and about 125 g/l sodium hydroxide (NaOH), (Rakhshani et al.1987, Rakhshani & Varghese, 1987). Cupric sulphate was dissolved first in distilled water giving it a light blue color. Then lactic acid was added. Finally, a sodium hydroxide solution was added, changing the color of the solution to dark blue with pH = 9. A copper clad for printed circuit board, with dimension 50 μm, 2.5 × 7 cm², was used as the anode. Copper clad and conducting glass slides coated with ITO and SnO_2 were used as a cathode. Experience shows that impurities (such as dirt, finger prints, etc.) on the starting surface material have a significant impact on the quality of the cuprous oxide. Therefore, mechanical and chemical cleaning of the electrodes, prior to the cell preparation, is essential. Copper boards were polished with fine emery paper. After that, they were washed by liquid detergent and distilled water. The ITO substrates were washed by liquid detergent and rinsed with distilled water. The SnO_2 substrates were soaked in chromsulphuric acid for a few hours and rinsed with distilled water. Before using all of them were dried.

Thin films of Cu_2O were electrodeposited by cathodic reduction of an alkaline cupric lactate solution at 60^0 C . The deposition was carried out in the constant current density regime. The deposition parameters, as current density, voltage between the electrodes and deposition time were changed. The Cu_2O films were obtained under following conditions: 1) current density j = 1,26 mA/cm^2, voltage between the electrodes V = 0,3 - 0,38 V and deposition time t = 55 min. Close to the value of current density, deposition time and Faraday's law, the Cu_2O oxide layer thickness was estimated to be about 5 μm.

The potentiostatic mode was used for deposition the Cu_2O films on glass coated with SnO_2 prepared by spray pyrolisis method of 0.1 M water solution of $SnCl_2$ complexes by NH_4F . The applied potential difference between anode and cathode was constant. It was found that suitable value is V = 0,5 to 0,6 V. The deposition current density at the beginning was dependent on the surface resistance of the cathode. For a fixed value of the potential, the current decreased with increasing film thickness. The film thickness was dependent on deposition current density j. For current density of about 1 mA/cm^2 at the beginning and deposition time of about 2 h, the film thickness was 5-6 μm approximately. The thickness of deposited film was determined using a weighting method, as $d = m/\rho s$, where m is the mass and s is the surface of the film. A density ρ, of 5.9 g/cm^3 was used.

The deposition of Cu_2O on a commercial glass coated with ITO was carried out under constant current density. The ITO/Cu_2O films was obtained under the following conditions: current density j = 0,57 mA/cm^2, voltage between the electrodes V = 1,1 - 1,05 V and deposition time t = 135 min. The Cu_2O oxide layer thickness was estimated to be about 5 μm. All deposited films had reddish to reddish-gray color.

2.1.2 Structural properties

The structure of the films was studied by X – ray diffraction, using CuKα radiation with a wavelength of 0.154 nm. The Bragg angle of 2θ was varied between 20^0 and 50^0. The XRD spectrums of the films samples, deposited on copper, glass coated by SnO_2 and glass coated by ITO are shown in Fig.1, Fig.2 and Fig.3 respectively. It was found that all films are polycrystalline and chemically pure Cu_2O with no traces of CuO. XRD peaks corresponded to Cu_2O and the substrate material. The XRD spectrums indicate a strong Cu_2O peak with (200) preferential orientation.

2.1.3 Morphological properties

The surface morphology of the films was studied by a scanning electron microscope JEOL model JSM 35 CF. Fig.4, Fig.5 and Fig.6 show the scanning electron micrographs of Cu_2O films deposited on copper, glass coated by SnO_2 and glass coated by ITO respectively. The photographs indicate a polycrystalline structure. The grains are very similar to each other in size and in shape. They are about 1 μm and less in size for the film deposited on copper, 1-2 μm for the film deposited on SnO_2 and about 1 μm for the film deposited on ITO.

2.1.4 Optical band-gap energy determination

The optical band-gap is an essential parameter for semiconductor material, especially in photovoltaic conversion. In this work it was determined using the transmittance spectrums of the films. The optical transmission spectrums were recording on Hewlett-Packard (model 8452 A) spectrophotometer in the spectral range 350-800 nm wavelength. Thin layers of a transparent Cu_2O were preparing for the optical transmission spectrums recording. The optical transmission spectrum of about 1,5 μm thick Cu_2O film deposited on glass coated with SnO_2 is presented in Fig.7. There are two curves, one (1) recorded before annealing and the other one (2) after annealing of the film for 3h at 130^{0}C.

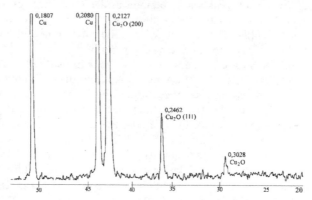

Fig. 1. X-ray diffraction spectrum of a Cu$_2$O film deposited on copper

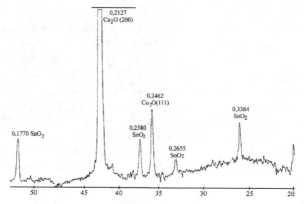

Fig. 2. X-ray diffraction spectrum of a Cu$_2$O film deposited on SnO$_2$

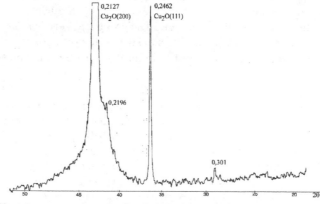

Fig. 3. X-ray diffraction spectrum of a Cu$_2$O film deposited on ITO

Fig. 4. Micrograph obtained from a scanning electron microscope of Cu$_2$O deposited on copper

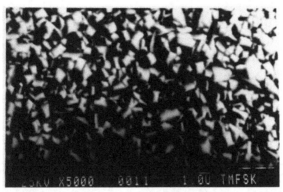

Fig. 5. Micrograph obtained from a scanning electron microscope of Cu$_2$O deposited on SnO

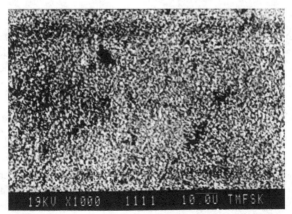

Fig. 6. Micrograph obtained from a scanning electron microscope of Cu$_2$O deposited on ITO

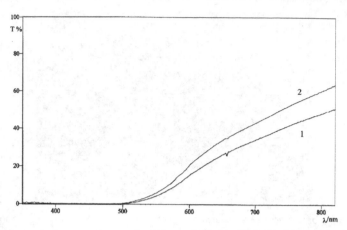

Fig. 7. Optical transmission spectrum of a 1,5 µm thick Cu_2O/SnO_2 film

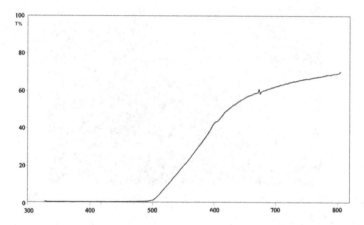

Fig. 8. Optical transmission spectrum of a 0,9 µm thick Cu_2O/ITO film

We can see that there is no difference in the spectrums. The absorption boundary is unchangeable. That means that the band gap energy is unchangeable with or without annealing. The little difference comes from different points recording, because the thickness of the film is not uniform. The transmittance spectrum of about 0,9 µm thick Cu_2O film, deposited on ITO, is presented in Fig. 8.

For determination of the optical band gap energy E_g , the method based on the relation

$$\alpha h\nu = A(h\nu - E_g)^{n/2},\tag{1}$$

has been used, where n is a number that depends on the nature of the transition. In this case its value was found to be 1 (which corresponds to direct band to band transition) because that value of n yields the best linear graph of $(\alpha h\nu)^2$ versus $h\nu$.

The values of the absorption coefficient α were calculated from the equation

$$\alpha = \frac{A}{d},$$ (2)

where d is the film's thickness determined using weighing method, and A is the absorbance determined from the values of transmittance, $T(\%)$, using the equation

$$A = \ln\frac{100}{T(\%)}.$$ (3)

The values of the optical absorption coefficient α in dependence on wavelength are shown in Fig. 9 for Cu_2O/SnO_2 film and Fig. 10 for Cu_2O/ITO film

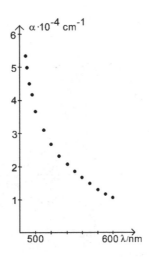

Fig. 9. Coefficient α vs wavelength λ for Cu_2O/SnO_2 film

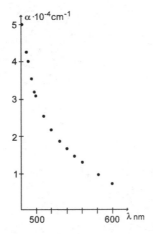

Fig. 10. Coefficient α vs wavelength λ for Cu_2O/ITO film

Fig.11 and Fig.12 show $(\alpha h\nu)^2$ versus $h\nu$ dependence for the Cu_2O/SnO_2 film and Cu_2O/ITO film corresponding. The intersection of the straight line with the $h\nu$ axis determines the optical band gap energy E_g. It was found to be 2,33 eV for Cu_2O/SnO_2 film and 2,38 eV for Cu_2O/ITO. They are higher than the value of 2 eV given in the literature and obtained for Cu_2O polycrystals. These values are in good agreement with band gaps

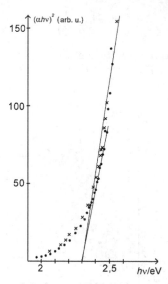

Fig. 11. Graphical determination of the optical band gap energy for Cu_2O/SnO_2 film

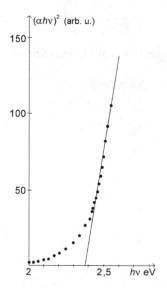

(x - before annealing; · - after annealing)

Fig. 12. Graphical determination of the optical band gap energy for Cu_2O/ITO film

determined from the spectral characteristics of the cells made with electrodeposited Cu_2O films. The value of the energy band gap of Cu_2O/ITO is little higher than the value of Cu_2O/SnO_2 film. The reason is maybe different size of the grains.

Fig.11 shows that there is no different in optical band gap energy determined from the curve plotted before annealing and from the curve plotted after annealing. Also, Fig.11 and Fig.12 show that there is no shape absorption boundary in the small energy range of the photons. Probably defects and structural irregularities are present in the films.

The optical band-gap of the films was determined using the transmitance spectrums. It was found to be 2,33 eV for Cu_2O/SnO_2 film and 2,38 eV for Cu_2O/ITO.

3. Preparation of the Cu_2O Schottky barrier solar cells

Cu_2O Schottky barrier solar cells can be fabricated in two configurations, the so called back wall and front wall structures. By vacuum evaporating a thin layer of nickel on the Cu_2O film, photovoltaic cells have been completed as back wall type cells (Fig.13), or by depositing carbon or silver paste on the rear of the Cu_2O layers, photovoltaic cells have been completed as front wall type cells (Fig.14). Nickel, carbon or silver paste are utilized to form ohmic contacts with cuprous oxide films. From the energy band diagram (Fig.15) we can see that the Cu_2O work function $\Phi_s = \chi + 1,7$ eV, (χ is the electron affinity of Cu_2O) (Olsen et al.,1982, Papadimitriou et al.,1990). That means that Cu_2O will make ohmic contact with metals characterized with work function higher than 4,9 eV, as are Ni, C. Gold and silver essentially form ohmic contacts. A carbon or silver back contact was chosen because of simplicity and economy of the cell preparation. The rectifying junction exists at the interface between the cooper and Cu_2O layers in the case of back wall cells. In the case of front wall cells the rectifying junction exists at the interface between the SnO_2 (ITO) and Cu_2O layers.

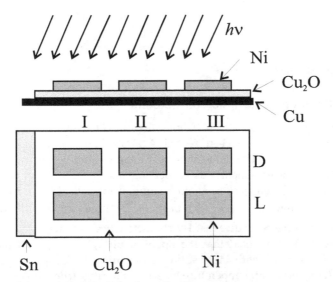

Fig. 13. Profile and face of Cu/Cu_2O back wall cell structure

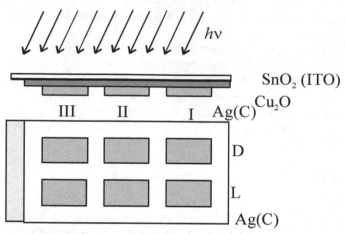

Fig. 14. Profile and rare of SnO_2 (ITO)/Cu_2O front wall cell structure

The evaporation of nickel has been made with Balzers apparatus under about $5,33 \times 10^{-3}$ Pa pressure. The optical transmission of the nickel layer was 50% for 550nm wavelength. The total cell active area is 1.0 cm². Antireflectance coating or any special collection grids have not been deposited.

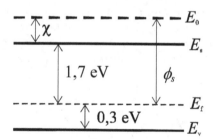

Fig. 15. Energy band diagram for Cu_2O

4. Current-voltage characteristics of the cells

The current-voltage characteristics of the best ITO/Cu_2O/C, Ni/Cu_2O/Cu and SnO_2/Cu_2O/C solar cells have been recorded in darkness and under 100 mW/cm² illumination, point by point. The light intensity was measured by Solar Meter Mod.776 of Dodge Products. The measurement was carried out using an artificial light source with additional glass filter, 10 mm thick to avoid heating of the cells. I-V characteristics, Fig.16, Fig.17 and Fig.18, were recorded first with periodically illumination of the source (curve ○) to avoid the heating of the cell. After that I-V characteristics were recorded with continually illumination (curve ×). It is noted that the open circuit voltage V_{oc} and the short circuit current density I_{sc} decrease with increase in temperature. V_{oc} drops because of increase reverse current saturation with temperature because minority carriers increase with increase in temperature. I_{sc} decrease because of increase the recombination of the charges.

It should be stressed that this cells showed photovoltaic properties after heat treatment of the films for 3 hrs at 130 °C in a furnace. This possibly results in a decrease of sheet resistance value of the Cu_2O films, which was not measured, or in transformation the Cu_2O semiconductor from n to p type after heat treatment. Before heating V_{oc} and I_{sc} were about zero or negative. The serial resistance R_s and shunt resistance R_{sh} for all types of the cells were evaluated from I-V characteristics.

Cell type	R_{so} kΩ	R_s kΩ	R_{sh} kΩ
ITO/Cu_2O/C	10	1,02	76
Ni/Cu_2O/Cu	20	8,3	40
SnO_2/Cu_2O/C	14	3,3	25

Table 1. Serial and shunt resistance

The values are given in Table 1. R_{so} is evaluated from the dark characteristics (curve Δ) as dV/dI for higher values of forward applied voltage. R_{sh} is evaluated as dV/dI from the dark characteristics in reverse direction for lower values of the applied voltage (Olsen & Bohara, 1975). R_s is evaluated from the light I-V characteristics and it decreases with illumination. That means that R_s is photoresponse. The high series resistance R_s and low shunt resistance R_{sh} are one of the reasons for poor performance of the cell.

Several cell parameters were evaluated from the I-V characteristics. Table 2 contains the optimal current and voltage values (I_m and V_m), the open circuit voltage (V_{oc}), the short circuit current (I_{sc}) and evaluated values of the fill factor FF $\left(FF = \dfrac{I_m V_m}{I_{sc} V_{oc}} \right)$, the efficiency η $\left(\eta = FF \dfrac{I_{sc} V_{oc}}{P_{in}} \right)$ and diode factor n.

Cell type	I_m μA	V_m mV	I_{sc} μA	V_{oc} mV	FF %	η 10^{-2} %	n
ITO/Cu_2O/C	130	180	245	340	28	2,34	2,23
Ni/Cu_2O/Cu	28	120	50	270	24	0,70	2,06
SnO_2/Cu_2O/C	46	90	74	225	25	0,41	2,20

Table 2. Cell parameters

The diode factor was evaluated from the logarithmic plot of the dependence of I_{sc} versus V_{oc} which were measured for different illumination. The diode factor defined as

$$n = \frac{q}{kT} \frac{\Delta V_{oc}}{\Delta \ln I_{sc}} \qquad (4)$$

is about 2 for all type of the cells.

The performances of the cells depend on the starting surface material, the type of the junction, post deposition treatment and the ohmic contact material. From the I-V characteristics, we can see that the cells are with poor performances, low fill factor FF and very low efficiency. The high R_s and low R_{sh} (which is very far from ideally solar cell) are one of the reasons for poor performances. Because of high series resistance R_s, the values of the short circuit current density are very low. By depositing gold instead of nickel or graphite paste, the performance may be improved by decreasing of R_s.

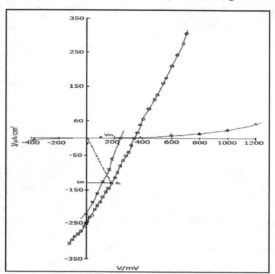

Fig. 16. I-V characteristics for ITO/Cu$_2$O/C solar cell ∘-periodically illumination (100 mW/cm²); ×-continually illumination(100 mW/cm²); Δ-dark characteristic

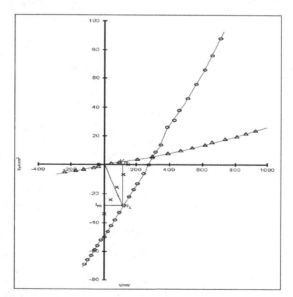

Fig. 17. I-V characteristics for Ni/Cu$_2$O/Cu solar cell o-periodically illumination (100 mW/cm²); ×-continually illumination(100 mW/cm²); Δ-dark characteristic

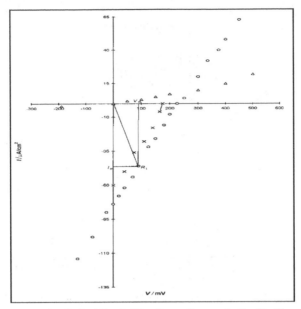

Fig. 18. I-V characteristics for $SnO_2/Cu_2O/C$ solar cell o-periodically illumination (100 mW/cm²); ×-continually illumination(100 mW/cm²); Δ-dark characteristic

5. Potential barrier height determination of the cells

Capacitance as a function of reverse bias voltage at room temperature of $Ni/Cu_2O/Cu$, $SnO_2/Cu_2O/$graphite and $ITO/Cu_2O/$graphite solar cells was measured by RCL bridge on alternating current (HP type) with built source with 1000 Hz frequency.

Results for $1/C^2$ versus reverse bias voltage for all these types of cells are shown in Fig 19, Fig 20 and Fig 21, before annealing (□), immediately after annealing (•) and after three months of annealing (×). The dependence is straight line. The intercepts of the straight line with x-axis correspond to the barrier height V_b. Cu/Cu_2O cell showed photovoltaic effect without post deposition heat treatment and their photovoltaic properties are almost unchangeable in time (fig.19). In contrast to this cell, the ITO/Cu_2O (fig.20) and SnO_2/Cu_2O (fig.21) cells no showed photovoltaic properties and no potential barrier was found to exist (Georgieva &Ristov, 2002). Before annealing, the open circuit voltage V_{oc} and the short circuit current I_{sc} were about zero.

After annealing of the films for 3 h at 130⁰C, the devices exhibited good PV properties and the potential barrier excised. But this situation was not stationary. That is another essential factor in the properties of these cells indicating the possibility of chemical changes in ITO/Cu_2O and SnO_2/Cu_2O junction (Papadimitriou et al.,1981).

The values of barrier height V_b and the open circuit voltage V_{oc} upon illumination by an artificial white light source of 100 mW/cm² for all types of cells are presented in table 3. Also in this table are given their values after aging for 3 months (*). Only Cu/Cu_2O cell has stationary values of V_b and V_{oc}. The values of barrier height V_b are great then the values of open circuit voltage V_{oc}. The great V_b gives the great V_{oc}, in consent with the photovoltaic theory.

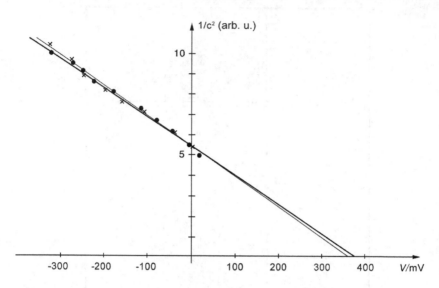

Evaluation of the barrier height, before annealing (□); after annealing (•); after 3 months of annealing (×).

Fig. 19. 1/C² vs applied voltage of Cu/Cu₂O cell.

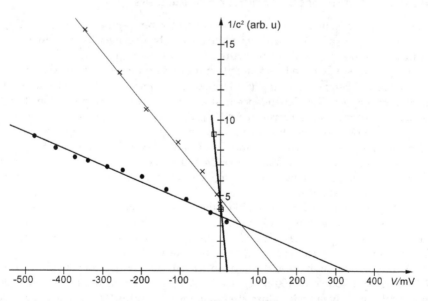

Evaluation of the barrier height, before annealing (□); after annealing (•); after 3 months of annealing (×).

Fig. 20. 1/C² vs applied voltage of ITO/Cu₂O cell.

Cell type	Cu/Cu$_2$O	ITO/Cu$_2$O	SnO$_2$/Cu$_2$O	
V_b (mV)	(mV)	378	330	180
V_{oc} (mV)	310	249	118	
V_b^* (mV)	370	150	60	
V_{oc}^* (mV)	310	105	30	

Table 3. Values of barrier height V_b and open circuit voltage V_{oc} for all types of cells after annealing and after aging for 3 months (*).

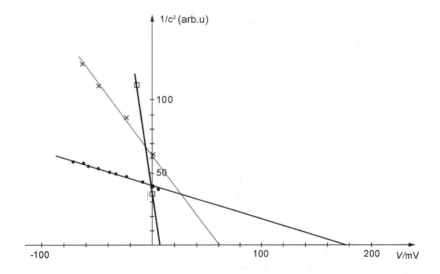

Evaluation of the barrier height, before annealing (□); after annealing (•); after 3 months of annealing (×).

Fig. 21. 1/C^2 vs applied voltage of SnO$_2$/Cu$_2$O cell.

6. ZnO/Cu$_2$O heterojunction solar cells

Until now, we have made Schottky barrier solar cells. As we could not improve their efficiency and their stability, we decided to make heterojunction p-n solar cells based on a p-type Cu$_2$O thin films. We selected ZnO as an n-type semiconductor. ZnO is a transparent oxide that is widely used in many different applications, including thin film solar cells. The p-n junction was fabricated by potentiostatic deposition of the ZnO layer onto SnO$_2$ conducting glass with a sheet resistance of 14 Ω/ and potentiostatic deposition of Cu$_2$O onto ZnO, Fig.22.

6.1 Electrochemical depositing of ZnO

ZnO/Cu$_2$O heterojunction solar cells were made by consecutive cathodic electrodeposition of ZnO and Cu$_2$O onto tin oxide covered glass substrates. Zinc oxide (ZnO) was cathodically deposited on a conductive glass substrate covered with SnO$_2$ as cathode by a potentiostatic method (Dalchiele et al.,2001, Izaki et al.,1998, Ng-Cheng-Chin et al.,1998). Conducting glass slides coated with SnO$_2$ films are commercial samples. The electrolysis takes place in a

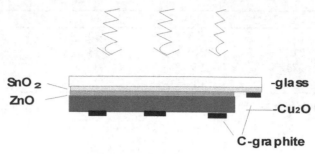

Fig. 22. Profil of ZnO/Cu_2O hetrojunction solar cells

simple aqueous 0,1M zinc nitrate $[Zn\,(NO_3)_2]$ solution with pH about 6, maintained at 70ºC temperature. The cathodic process possibly can be described by the following reaction equations (Izaki & Omi, 1992):

$$Zn(NO_3)_2 \to Zn^{2+} + 2NO_3^-$$

$$NO_3^- + H_2O + 2e^- \to NO_2 + 2OH^-$$

$$Zn^{2+} + 2OH^- \to Zn(OH)_2 \to ZnO + H_2O \qquad (5)$$

ZnO films were electrochemically grown at constant potential of 0.8 V between the anode and cathode. For a fixed value of the potential, a current density decreased with increasing the film thickness. The deposition time was varying from 10 min to 30 min. Deposited films were rinsed thoroughly in distilled water and allowed to dry in air at room temperature. The anode was zinc of 99.99% purity.

The deposition conditions of the thin films of Cu_2O have been described in 2.1.1. The deposition potential is pH sensitive. It suggests, also and it has already been reported that the Cu_2O layer was formed by the following reaction:

$$2Cu^{2+} + 2e^- + 2OH^- \to Cu_2O + H_2O, \qquad (6)$$

even this reaction does not explain the large pH dependence of deposition potential (Izaki et al. 2007, Wang & Tao, 2007). The present study was conducted, in a first instance, on undoped zinc oxide films and cuprous (I) oxide films. The structure of the films was studied by X-ray diffraction measurements using monochromatic Cu K_α radiation with a wavelength of 0,154 nm operated at 35 kV and 24 mA. Morphology and grain size was determined through micrographs on a JEOL JSM 6460 LV scanning electron microscope.

Figure 23 shows the X-ray diffraction patterns of ZnO film prepared at 0.8 V potential for 10 min. The Bragg angle of 2θ was varied between 20^0 and 70^0. It can be seen that the film has crystalline structure. XRD peaks corresponding to ZnO (signed as C) and the substrate material SnO_2 (signed as K) were determined with JCPDS patterns. The XRD spectrum indicates a strong ZnO peak with a (0002) or (1011) preferential orientation.

Figure 24 shows a scanning electron micrograph of undoped electrodeposited ZnO film.The photograph shows small rounded grains. It is difficult to determine the grain size from the micrograph. But using Scherrer's equation ($D = \dfrac{0,9\lambda}{\beta\cos\theta}$), the apparent crystallite size of ZnO is about 20nm, which means that it is nanostructured film

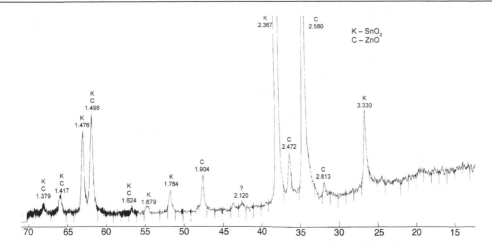

Fig. 23. X-ray diffraction spectrum of undoped electrodeposited ZnO film at 65⁰C

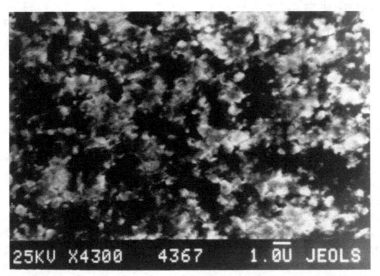

Fig. 24. SEM micrograph of undoped electrodeposited pure ZnO

Thin films of ZnO grown by electrochemical deposition technique on SnO_2/glass substrate are optically transparent in a visible spectral region, extending to 300 nm wavelength.

The transmission is relatively low (~ 50%) in the blue region (400–450 nm) Fig.25. The transmission maximum is about 60–70% through the red light region. Probably defects and structural irregularities are presented in the films, indicating low transmission.

Assuming an absorption coefficient α corresponding to a direct band to band transition and making a plot of $(\alpha h v)^2$ versus energy $h v$, the optical band gap energy Eg was determined through a linear fit. It was found to be 3.4 eV , which corresponds to the documented room temperature value of 3.2 to 3.4 eV.

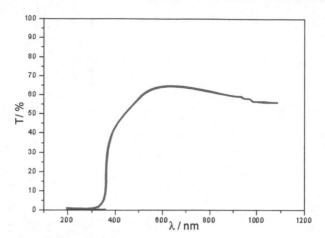

Fig. 25. Optical transmission spectrum of ZnO film

6.2 Some characteristic of the cells

To complete $Cu_2O/ZnO/SnO_2$ heterojunction as solar cell, thin layer of carbon paste or carbon spray was deposited on the rear of the Cu_2O. Front wall cells were formed. A carbon back contact was chosen because of simplicity and economy of the cell preparation and because the cells with carbon give high values of the short circuit current density despite the evaporated layer of nickel. The total cell active area was 1 cm². Antireflectance coating or any special collection grids have not been deposited. The best values of the open circuit voltage $V_{oc}= 330$ mV and the short circuit current density $I_{sc} = 400$ μm/cm² were obtained by depositing carbon paste and illumination of 100 mW/cm². The V_{oc} increases as logarithmic function with solar radiation, ($Voc = \dfrac{kT}{e}\ln\left(\dfrac{I_{sc}}{I_0}+1\right)$). The Isc increases linear with solar radiation, (Fig.26).

Our investigations show that the ZnO layer improves the stability of the cells. That results in a device with better performances despite of the Schhotky barrier solar cells (Cu_2O/SnO_2). First, the cells show photovoltaic properties without annealing, because potential barrier was formed without annealing. The barrier fell for a few days which result in decreasing the open circuit voltage despite the values of V_{oc} for just made cells. It decreases from 330 mV to 240 mV. But after that the values of V_{oc} keep stabilized, because of stabilized barrier potential. It wasn't case with Schotky barrier solar cells, because barrier potential height decreases with aging. In ZnO/Cu_2O cells, thermal equilibrium exists. The V_{oc} decreases and I_{sc} increases with increasing the temperature, that is characteristic for the real solar cell. It could be seen from the current-voltage (I-V) characteristic in incident light of 50mW/cm², Fig.27.

Barrier potential height was determined for one device from capacitance measurement as a function of reverse bias voltage at room temperature. Capacitance dependence of reverse bias voltage at room temperature was measured by RCL bridge on alternating current (HP type) with bilt source with 1000 Hz frequency. Results for ($1/C^2$) versus voltage are shown in Figure 28. The Cu_2O/SnO_2 cells without the ZnO layer show a lower V_{oc}. The improvement in V_{oc} could be due to the increase of the barrier height using ZnO layer as n-type semiconductor.

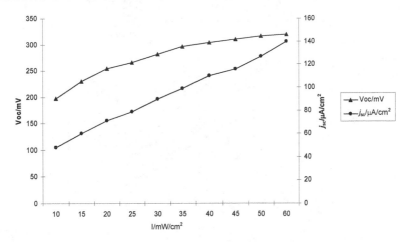

Fig. 26. Dependence of the V_{oc} and I_{sc} vs. solar irradiation

The values of barrier height V_b and the open circuit voltage V_{oc} upon illumination of 100 mW/cm² for just made cell and the cell after few days are presented in table 4. Also in this table are given their values after few days of depositing. The values of the barrier height are great than the values of open circuit voltage V_{oc}. The grate V_b gives the great V_{oc}, that correspondent to the photovoltaic theory.

$Cu_2O/ZnO/SnO_2$ cell	V_b(mV)	V_{oc}(mV)
just made	368	330
after few days	276	240

Table 4. Values of barrier height V_b and open circuit voltage V_{oc} for just made cell and after few days.

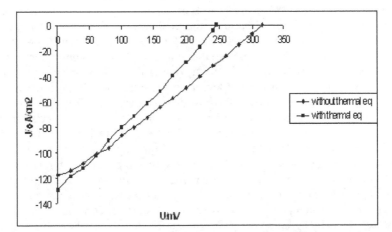

Fig. 27. Volt-current characteristics of the $Cu_2O/ZnO/SnO_2$ solar cell upon 50mW/cm² Illumination

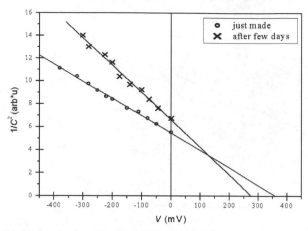

Fig. 28. $1/C^2$ vs applied voltage of $Cu_2O/ZnO/SnO_2$ cell

The values Voc=316 mV, Isc=0,117 mA/cm^2, fill factor =0,277, upon 50mW/cm^2 illumination are compared with the values: Voc=190 mV, Isc=2,08 mA/cm^2, fill factor = 0,295; upon 120mW/cm^2 illumination (Katayama et al. 2004) made with electrochemical deposition technique. Maybe doping of the ZnO films with In, Ga and Al (Machado et al., 2005, Kemell et al., 2003) will decrease the resistivity and increase the electro conductivity of the films, consequently and the short circuit current density of the cells.

7. Conclusion

The performance of the Cu_2O Schottky barrier solar cells are found to be dependent on the starting surface material, the type of the junction, post deposition treatment and the ohmic contact material. Better solar cells have been made using an heterojunction between Cu_2O and n-type TCO of ZnO. It is a suitable partner since it has a fairly low work function. Our investigation shows that the ZnO layer improves the stability of the cells. That results in a device with better performances despite of the Schhotky barrier solar cells (Cu_2O/SnO_2). First, the cells show photovoltaic properties without annealing, because potential barrier was formed without annealing. To improve the quality of the cells, consequently to improve the efficiency of the cells, it has to work on improving the quality of ZnO and Cu_2O films, because they have very high resistivity, a factor which limits the cells performances. Doping of the ZnO films with In, Ga and Al will decrease the resistivity of the deposited films and increase their electroconductivity. SEM micrographs show that same defects are present in the films which act as recombination centers. Behind the ohmic contact, maybe one of the reason for low photocurrent is just recombination of the carriers and decreasing of the hole cocentracion with the time. The transmittivity in a visible region have to increase. Also, it is necessary to improve the ohmic contact, consequently to increase the short circuit current density (Isc). For further improvement of the performances of the cells maybe inserting of a buffer layer at the heterojunction between Cu_2O and ZnO films will improve the performance of the cells by eliminating the mismatch defects which act as recombination centers. Also it will be protection of reduction processes that maybe exists between ZnO and Cu_2O.

Even low efficiency it may be acceptable in countries where the other alternative energy sources are much more expensive.

8. Acknowledgment

Part of this work has been performed within the EC funded RISE project (FP6-INCO-509161). The authors want to thank the EC for partially funding this project.

9. References

Dalchiele E.A., Giorgi P.,.Marotti R.E, at all. Electrodeposition of ZnO thin films on n-Si(100), *Solar Energy Materials &Solar Cells* 70 (2001) 245-254

GeorgievaV. Ristov M. (2002) Electrodeposited cuprous oxide on indium tin oxide for solar applications, *Solar Energy Materials & Solar Cells* 73, p 67-73

Izaki M., Omi T,J. (1992).,*Electrochem.Soc.*1392014

Izaki M, Ishizaki H., Ashida A. at all. (1998), *J.Japan Inst.Metals*, Vol. 62,No.11pp.1063-1068

Izaki M., Shinagawa T., Mizuno K., Ida Y., Inaba M., and Tasaka A., (2007) Electrochemically constructed p-Cu_2O/n-ZnO heterojunction diode for photovoltaic device *J.Phys.D:Appl.Phys.*40 3326-3329.

Jayanetti J.K.D., Dharmadasa I.M, (1996), *Solar Energ.Mat.andSolar Cells* 44 251-260

Katayama J., Ito K. Matsuoka M. and Tamaki J., (2004) Performance of Cu_2O/ZnO solar cells prepared by two-step electrodeposition *Journal of Applied Electrochemistry*, 34: 687-692,

Kemell M., Dartigues F., Ritala M., Leskela M, (2003) Electrochemical preparation of In and Al doped Zno thin films for $CuInSe_2$ solar cells, *Thin Solid Films* 434 20-23

Machado G., Guerra D.N., Leinen D., Ramos-Barrado J.R. Marotti R.E., Dalchiele E.A., (2005), Idium doped zinc oxide thin films obtained by electrodeposition ,*Thin Solid Films* 490 124-131

Minami T.,.Tanaka H, Shimakawa T., Miyata T., Sato H., (2004) High-Efficiency Oxide Heterojunction Solar cells Using Cu_2O Sheets *Jap.J.Appl.Phys.*,43, p.917-919

Mukhopadhyay A.K.,.Chakraborty A.K, Chattarjae A.P. and.Lahriri S.K, (1992), *Thin Solid Films*, 209, 92-96

Olsen L.C., Bohara R.C., Urie M.W., (1979) *Appl.Phys.Lett,* 34, p. 47

Olsen L.C., Addis F.W. and Miller W., (1982-1983), *Solar Cells*, 7 247-249

Papadimitriou L., Economu N.A and Trivich, (1981), *Solar Cells*, 3 73

Papadimitriou L., Valassiades O. and Kipridou A., (1990), *Proceeding*, 20th ICPS, Thessaloniki, 415-418

Ng-Cheng-Chin F., Roslin M.,.Gu Z.H and Fahidy T.Z., (1998) *J.Phys.D.Appl.Phys.*31 L71-L7

Rai B.P., (1988) Cu_2O Solar Cells *Sol. Cells* 25 p.265.

Rakhshani A.E (1986) Preparation, characteristics and photovoltaic properties of cuprous oxide-A review *Solid-State Electronics* Vol. 29.No.1. pp.7-17.

Rakhshani A.E., Jassar A.A.Al and.Varghese J. (1987) Electrodeposition and characterization of cuprous oxide *Thin Solid Films*, 148,pp.191-201

Rakhshani A.E.and Varghese J., (1987) Galvanostatic deposition of thin films of cuprous oxide *Solar Energy Materials*, 15,23,

Rakhshani A.E., Makdisi Y. and Mathew X., (1996), *Thin Solid Films*, 288, 69-75

Stareck, U.S. Patents 2, 081, 121 *Decorating Metals*, 1937

Wang L. and Tao M. (2007) Fabrication and Characterization of p-n Homojunctions in Cuprous Oxide by Electrochemical Deposition, *Electrochemical and Solid-State Letters*, 10 (9) H248-H250

3

Enhanced Diffuse Reflection of Light by Using a Periodically Textured Stainless Steel Substrate

Shuo-Jen Lee and Wen-Cheng Ke
Yuan Ze University, Taiwan,
R.O.C.

1. Introduction

The flexible solar cells fabricated on a stainless steel substrate are being widely used for the building of integrated photovoltaics (BIPVs) in recent years. Because stainless steel has many advantages, such as low cost, high extension, ease of preparing etc. It was believed that the wide application of BIPVs especially rooftop applications, would be the biggest market for flexible PV technology (Kang et al. 2006, Otte et al. 2006, Chau et al. 2010, Fung et al. 2008). Until now, one of the main challenges of the BIPVs remains how to improve the conversion efficiency. Since, the path length of the photovoltaic effect is considerable shorter in a thin film solar cell resulting in reduced efficiency. Many researchers have focused on light trapping, and have adopted a different TCO technology, such as LP-CVD, PVD, to increase the path length of the incoming light, and improve the photovoltaic conversion efficiency of thin film solar cells (Selvan et al. 2006, Llopis et al. 2005, Söderström et al. 2008, Müller et al. 2004). Moreover, light trapping provides some significant advantages including, reduction of the cell thickness, reduced processing time and reduced cost, improved cell efficiency and the improved stability of amorphous Si (a-Si:H).

The idea of trapping light inside a semiconductor by total internal reflection was reported by John in 1965 (John 1965). It also indicated that the effective absorption of a textured semiconductor film could be enhanced by as much as a factor of 60 over a plane-parallel film (Yablonovitch and Cody 1982). It should be mentioned that a major limitation to thin film solar cell efficiency is the long absorption length of the long wavelength photons and the low thickness of the absorber layer. The absorption length of amorphous silicon (a-Si:H) with a bandgap of 1.6 eV, for red and infrared solar photons, exceeds 1 μm and 100 μm, respectively (Ferlanto et al. 2002, Zhou and Biswas 2008). However, for a-Si:H the hole diffusion length is ~300-400 nm, which limits the solar cell absorber layer thickness to less than the hole diffusion length (Curtin et al., 2009). This makes it exceedingly difficult to harvest these photons since the absorber thickness of a p-i-n single junction solar cell is limited to only a few hundred nanometers for efficient carrier collection. In addition, the low-cost approach of thin-film silicon solar cells is very sensitive to film thickness, since the throughput increases with the decrease in layer thickness. Thus, sophisticated light trapping is an essential requirement for the design of thin-film solar cells (Rech et al., 2002).

Enhanced light-trapping in thin film solar cells is typically achieved by a textured metal backreflector that scatters light within the absorbing layer and increases the optical path length of the solar photons. In our recent researches [Lee et al., 2009], various processing techniques including, electro-polishing, sandblasting, photolithography, lift-off and wet-chemical etching were used to create periodically textured structures on the different types of stainless steel substrates. The relationships between the surface morphology of textured stainless steel substrate and optical properties will be carefully discussed.

2. Surface treatment of texturing stainless steel substrate

2.1 Electro-polishing process

In this study, electrochemical processing was used to achieve sub-micro texturing stainless steel substrate base on the fundamental electrochemical reaction items as (1)-(3).
Anode chemical reaction:

$$Fe^{2+}+2(OH)^-\rightarrow Fe(OH)_2 \text{ or } Fe(OH)_3 \qquad (1)$$

$$OH^-\rightarrow O_2\uparrow+H_2 \text{ (Parasitic reaction)} \qquad (2)$$

Cathode chemical reaction:

$$2H^+\rightarrow H_2\uparrow \qquad (3)$$

The electro-polishing system is shown in Fig. 1. The important parameters are as follows:
1. Substrate clean by acid-washing in $H_2O_2:H_2SO_4$=1:3 solution.
2. Electrolyte solution (Na_2SO_4) with concentration of 60-100 (g/L).
3. Current density in electro-polishing (EP) process is 0.1-1.0 (A/cm^2).

The clamp was used to hold the anode and cathode plates. The anode and cathode plates were separated by Teflon with thickness of 1 cm. Fig. 2 shows the optical microscopy (OM) images of 304 SS substrate with and without EP process. The average surface roughness (Ra) of 304 SS substrate increased from 0.045 μm to 0.197 μm after the EP process with current density of 1A/cm^2 in 10 min.

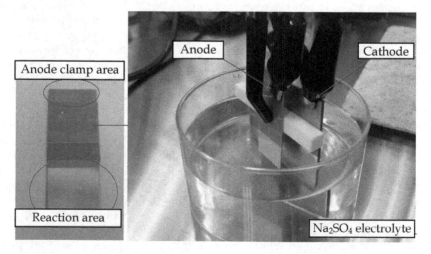

Fig. 1. Experimental set-up of the EP process.

2.2 Sand blasting process

The glass sand (#320) was used to form randomly textured surface with cave size of several µm to tens µm on the surface of stainless steel substrate. The average surface roughness (Ra) of 304 SS substrate increased from 0.277 µm to 6.535 µm after the sand blasting process. The OM images of raw 304 SS substrate and with sand blasting process were shown in Fig. 3.

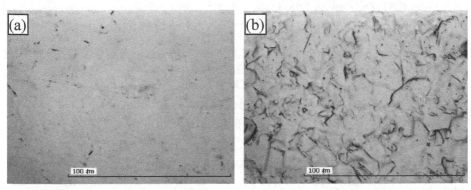

Fig. 2. The OM images (x2000) of (a) raw 304 SS substrate surface and (b) 304 SS substrate surface with EP process.

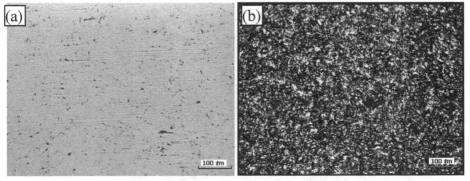

Fig. 3. The OM images (x400) of (a) raw 304 SS substrate surface and (b) 304 SS substrate surface with sand blasting process.

2.3 Photolithography process

The photo-mask patterns were designed by CAD. Photolithography is a process of using light to transfer a geometric pattern from a photo-mask to a photo-resist on a 430BA SS substrate. The steps involved in the photolithographic process are metal cleaning, barrier layer formation, photo-resist application, soft baking, mask alignment, exposure and development, and hard-baking. After the photolithographic process, the 430BA SS substrate is etched by aqua regia (HNO_3 : HCl=1 : 3). There are two types of photo-mask patterns: one, different diameters but with the same interval, and two, the same diameters but with a different interval. They are both designed to study light trapping for the application of thin film solar cells. Finally, silver coating technique by e-beam evaporation was used to improve the TR and DR rates of the 430BA SS substrate.

2.4 Lift-off and etching process

In this study, lift-off and etching processes were used to fabricate the different textures of the 304BA SS substrates. The striped texture was created on the 304BA SS substrate using the lift-off process. After the hard-baking process, a silver (Ag) thin film was deposited on the substrate by e-beam evaporation. An acetone solution was used to remove the residual photo resistor (PR). The depth of the striped texture was controlled by the thickness of the Ag thin film deposited. Four different striped textures were created on the 304BA SS substrates, including period/height: 6/0.1, 6/0.3, 12/0.1 and 12/0.3 μm. Two other types of textured 304BA SS substrate, the ridged-stripe and pyramid texture with 22.5 μm width were created by the etching process. After hard-baking, the 304BA SS substrate was etched by aqua regia (HNO_3 : HCl : DI water=1 : 3 : 4). The etching temperature was 28-35°C with an etching time of 7-12 min. to control the etching depth of the textured 304BA SS substrate. The detail experimental flow charts of lift-off and etching processes are shown in Fig. 4 and Fig. 5, respectively.

3. Optical properties of textured stainless steel substrate

3.1 Measurements of optical properties of textured stainless steel substrate

The total reflection (TR) and diffuse reflection (DR) rates of incident light from the textured substrate were carefully studied by using a Perkin Elmer Lambda 750S spectrometer. It was known that the specula reflection takes place on a smooth surface, and the angle of reflection is the same as the angle of incidence. DR is a phenomenon where an incident beam of light strikes an uneven or granular surface and then scatters in all directions. In Fig. 6, the 6 cm

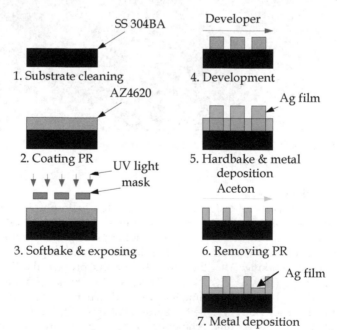

Fig. 4. The experimental flow charts of lift-off process.

Integrating Sphere is used for diffuse reflectance measurements. Reflectance measurements include total and diffuse reflectance at an incident angle of 8 degrees. Specular reflectance can be calculated from the total and diffuse reflectance measurements. The TR and DR rate of a textured substrate are important indexes when increasing the light trapping efficiency of thin-films solar cells.

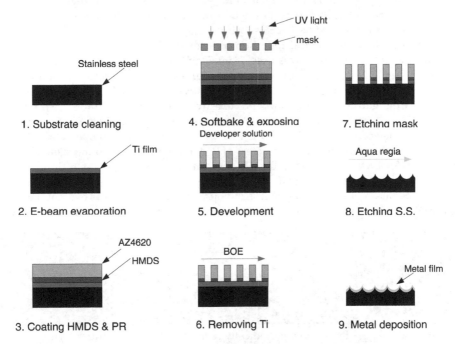

Fig. 5. The experimental flow charts of etching process.

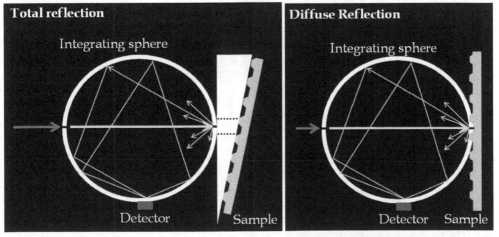

Fig. 6. The total reflection and diffuse reflection measured by integrating Sphere.

3.2 Optical properties on periodically textured 430BA stainless steel substrate

Lately the light trapping properties of textured substrates have attracted substantial interest because of their potential to reduce the thickness of solar cell material. In this study, the different kinds of textured patterns formed on 430BA SS substrate have been proposed for the purpose of trapping light in the application of thin film solar cells. Figure 7 shows the surface morphology of the 430BA SS substrate etched by using aqua regia. It should be noted that the dark and light regions of the OM images indicate the concave structure and the flat surface on the textured 430BA SS substrate, respectively.

In order to understand the optical reflection of a textured 430BA SS substrate, the Perkin Elmer Lambda 900 spectrometer was used to analyze both the TR and DR rates of incident light. The TR and DR rates versus the wavelength curves for the raw and textured 430BA SS substrates are shown in Fig. 8 and Fig. 9, respectively. The "D" and "G" indicate the diameter and the gap for these periodically textured 430BA SS substrates, respectively. It must be noted that the discontinued data line in the wavelength of 850 to 950nm was due to the change in detector, from a PMT to a PbS detector. First, we compared the textured 430BA SS substrates with different diameters of 2, 4 and 6μm and with the same interval of 3μm. In Fig. 8, it was found that the DR rate at the wavelength of 600nm increases substantially, from 4.5% of a raw 430BA SS substrate to 19.7 %, 23.1% and 31.8% for textured 430BA SS substrates with a diameter of 2, 4 and 6μm, respectively. It was evident that for the same areas of analysis, the larger the size of the concave shape, the worse the TR rate would

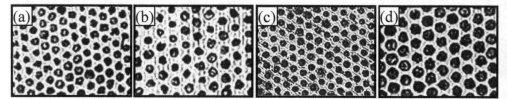

Fig. 7. The OM images of a concave periodically textured 430BA SS substrate with a diameter/gap of (a) 4/5 μm (b) 4/7 μm (c) 4/3 μm (d) 6/3 μm.

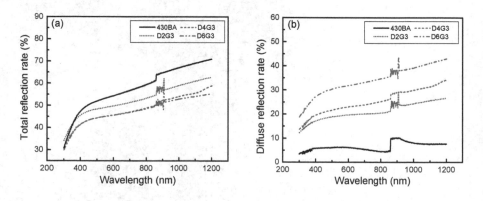

Fig. 8. The TR and DR rates versus the wavelength curves for raw 430BA SS substrate at different diameter of textured 430BA SS substrate.

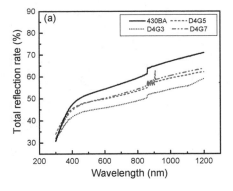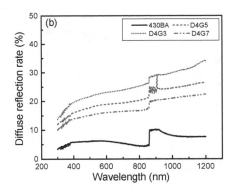

Fig. 9. The TR and DR rates versus the wavelength curves for raw 430BA SS substrate at different intervals of textured 430BA SS substrate.

be, resulting in a better diffuse reflection rate. We also investigated the TR and DR rates of the textured 430BA SS substrate with a different interval for samples with a fixed diameter of 4μm. In Fig. 9 we found that the DR rate at the wavelength of 600 nm decreased from 23.1% for a diameter/gap of 4/3 μm textured 430BA SS substrate to 18.9% and 16.1% respectively for a diameter/gap of 4/5μm and 4/7μm textured 430BA SS substrates. The decrease of the DR rate is related to the increase in the interval of the concave substrate.

The textured surface of the 430BA SS substrate leads to a lower TR rate compared to a specular surface of raw 430BA SS substrate. The lowering of the TR rate for the textured surface of the 430BA SS substrate can be understood on the basis of (a) the multiple scattering as a result of the multiple reflections from the textured surface of the 430BA SS substrate and a concomitant reduction in light intensity at each reflection due to the finite value of reflectance for 430BA SS, (b) light trapping in the indentations of a highly textured surface. Therefore, the results show that the textured 430BA SS substrate can generate a random distribution of light by reflection from a textured surface.

It is known that the incident light is reflected back into the cell for a second pass and subsequent passes. This phenomenon results in enhanced absorption in the cell. Thus, a back reflector should possess high reflectance in the solar part of the spectrum, which makes Ag a good candidate. Thus, we also performed the Ag coating on a textured 430BA SS substrate to study the TR and DR rates of incident light. The TR and DR rates versus the wavelength of textured 430BA SS with a silver film thickness of 300 nm are shown in Fig. 10. The peak at around 325 nm can be attributed to the diffuse reflectance spectrum of the deposited Ag film on the surface of the textured 430BA SS substrate (Xiong et al. 2003). In Fig. 10, the DR rate at the 600 nm wavelength are 40.6%, 47.2% and 64.6%, respectively for a diameter/gap of 2/3, 4/3 and 6/3 μm Ag film coated/textured 430BA SS substrate. The DR rate of Ag film coated/textured 430BA SS substrate increased about 2 times in comparison with the uncoated textured 430BA SS substrate (see Fig. 8). Similar results were also observed in Fig. 11 for the Ag film coated/textured 430BA SS substrate with a different interval for samples with a fixed diameter of 4μm. In addition, the TR rate increased to more than 90% for the Ag film coated/textured 430BA SS substrate which was an 80% improvement over the uncoated textured 430BA SS substrate.

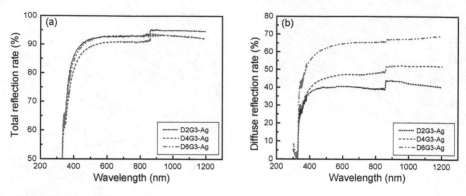

Fig. 10. The TR and DR rates versus wavelength curves for Ag film deposited at different diameter of textured 430BA SS substrate.

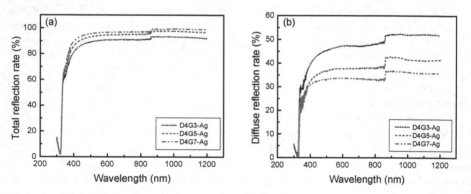

Fig. 11. The TR and DR rates versus wavelength curves for Ag film deposited at different intervals of textured 430BA SS substrate.

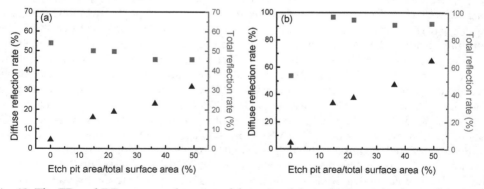

Fig. 12. The TR and DR rates as a function of the ratio of the etch pit area to the total surface area for (a) the textured 430BA SS and (b) the Ag film coated/textured 430BA SS.

From Fig. 8, it is evident that the TR and DR rates are not only dependent on the size of the concave shape but also depend on the interval of the concave substrate. Fig. 12(a) and (b) show the TR rate and DR rates as a function of the ratio of the etch pit area to the total surface area for the textured 430BA SS and the Ag film coated/textured 430BA SS, respectively. The ratio of the etch pit area to the total surface area is calculated by the number of pit in the total surface area multiplied by the single pit area divided by the total surface area. The total surface area is the analysis area in the spectrometer, measuring 1 cm². It is evident that the DR rate increased with the increase effectiveness of the pit regions compared to the smooth regions for both the textured 430BA SS and the Ag film coated/textured 430BA SS. However, the TR rate showed the opposite trend compared with the DR rate and decreased with the increase of the ratio of the etch pit area to total surface area. It is worth noting that once again the TR rate for the the Ag film coated/textured 430BA SS was more than 90% even the ratio of the etch pit area to the total surface area was 50%.

As shown in Figs. 8-11, we found that the increase in TR and DR rates as increase in light wavelength differed in the infrared range. The TR and DR rates clearly increased with the increasing light wavelength of the textured 430BA SS substrate. However, the TR and DR rates didn't increase with the increasing light wavelength of the Ag film coated/textured 430BA SS substrate. Huang et al. indicated that a metal with a lower work function can enhance the Raman signal of diamond films, which is referred to as surface-enhanced Raman scattering (SERS) (Huangbr et al. 2000). A very similar effect, surface-enhanced infrared absorption (SEIRA) was reported to occur with thin metal films (Hartstein et al. 1980, Hatta et al. 1982, Osawa 1997). Moreover, it was reported that the enhancement depends greatly on the morphology of the metal surface (Nishikawa et al. 1993). Fig. 13 shows the SEM image of an Ag film coated/textured 430BA SS substrate. It was found that the surface was covered with Ag particles ranging in size from tens to hundreds of nanometers. Thus, we believe that the difference in increase of the TR and DR rates in the infrared range for the textured 430BA SS and the Ag film coated/textured 430BA SS reflectors are due to the absorption in the infrared range by the Ag films. Further, the surface morphology was related to the thickness of the Ag film and must be carefully investigated in future study.

Fig. 13. The SEM image of Ag film coated/textured 430BA SS substrate.

3.3 Optical properties on periodically textured 304BA stainless steel substrate

Fig. 14 shows the OM images of the striped texture on the 304BA SS substrate. There are four patterns (i.e. the period/depth of 12/0.1, 12/0.3, 6/0.1, and 6/0.3 μm) which are designed and used to study the TR and DR rates of the 304BA SS substrate. The stripe width and depth of the samples were measured by a surface profiler. The TR and DR rates versus the wavelength curves for untreated and the stripe-textured 304BA SS substrates are shown in Fig. 15. The "P" and "D" indicate the period and the depth for the periodically textured 304BA SS substrates, respectively. It was found that the DR rate at the wavelength of 600 nm increased substantially, from 3.5% of an untreated 304BA SS substrate to 10.5%, 21.8%, 18.2% and 39.4% for textured 304BA SS substrates with a period/depth of 12/0.1, 12/0.3, 6/0.1, and 6/0.3 μm, respectively. In addition, the TR rate of the untreated 304BA SS was 67.7% and increased to ~97% for the striped textured 304BA SS substrate due to the high reflection of the Ag film on its surface. It was evident that for the same areas of analysis, the smaller the period and the larger the depth, the better the DR rate would be, resulting in a better diffuse reflection rate.

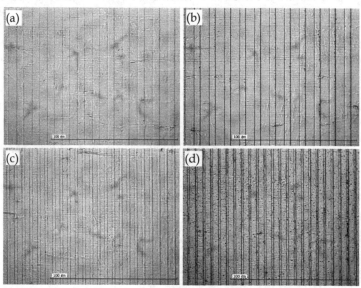

Fig. 14. The OM images of the stripe-textured 304BA SS substrate with a period/depth of (a) 12/0.1 μm (b) 12/0.3 μm (c) 6/0.1 μm (d) 6/0.3 μm.

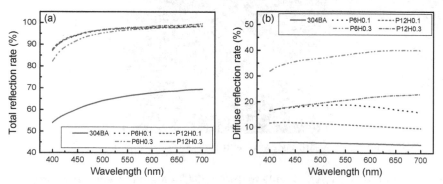

Fig. 15. The TR and DR rates versus the wavelength curves for untreated and stripe-textured 304BA SS substrates.

In our previous study (Lee et al. 2009), it was found that for a textured 430BA SS substrate the DR rate increased with the increased effectiveness of the etch-pit regions compared to that of the smooth regions. Thus, the large and deeply etched areas of the textured 304BA SS indicated that they can improve the DR rate of a textured 304BA SS substrate. In order to improve the DR rate even further, we design two other kinds of textured 304BA SS substrates, the ridged-stripe and the pyramid texture. 3D images of the ridged-stripe and pyramid texture are shown in Figs. 16(a) and (b), respectively. The etching depth and the width for both textured 304BA SS substrates were estimated to be ~6.5 μm and ~22.5 μm, respectively. The aspect ratio (i.e. depth/width) was ~1/3.5 indicating that the opening angle α of the textured surface was about ~120°. It should be noted that the etching depth is controlled by the PR thickness and the etching time. In general, a thick PR and a long etching time can create the deeper textured 304BA SS substrate.

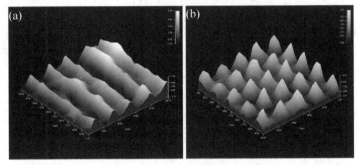

Fig. 16. The 3D images of (a) ridged-stripe and (b) pyramid 304BA SS substrate.

The TR and DR rates of the ridged-stripe and pyramid textured 304BA SS substrates are shown in Fig. 17. We found that the DR rate at the wavelength of 600 nm increased from 3.5 % for the untreated 304BA SS substrate to 60.1% for the pyramid and 63.1% for the ridged-stripe textured 304BA SS substrate. In addition, the DR rate also increased 1.5 times at the period/depth of 6/0.3 μm for the stripe-textured 304BA SS substrate. However, the textured substrates had a lower TR rate compared to the untreated 304BA SS substrate. The lowering of the TR rate for the textured surface of the 304BA SS substrate can be explained as follows (a) the multiple scattering is the result of the multiple reflections from the ridged-stripe or pyramid textured surface of the 304BA SS substrate, and the etching pit reduction in light intensity at each reflection is due to the finite value of the reflectance for the 304BA SS substrate, (b) light trapping occurs in the indentations of a highly textured surface. Therefore, the results show that the textured 304BA SS substrate can generate a random distribution of light through reflection from a textured surface.

It is well known that the incident light is reflected back into the cell for a second pass and subsequent passes. This phenomenon results in enhanced absorption in the cell. Thus, a back reflector must possess high reflectance in the solar part of the spectrum, making Ag or Al good candidates. However, Al films absorb the incident light wavelength of 800 nm and reduce the light conversion efficiency. On the other hand, the reflection of Ag film can achieve 99% from the visible to the IR wavelength (Jenkins and white 1957). Thus, we also used an Ag coating on a textured 304BA SS substrate to study the TR and DR rates of incident light. The TR and DR rates versus the wavelength of ridged-stripe and pyramid textured 304BA SS substrates with a silver film thickness of 150 nm are shown in Fig. 18. The

DR rates at the 600 nm wavelength were 95.6% and 96.8%, for the ridged-stripe and pyramid Ag film coated/texture 304BA SS substrates, respectively. The DR rate increased about 15-fold in comparison with the Ag coated untreated 304BA SS substrate. In addition, the TR rates at the 600 nm wavelength were 96.7% and 96.8%, for the ridged-stripe and pyramidal Ag film coated/texture 304BA SS substrates, respectively.

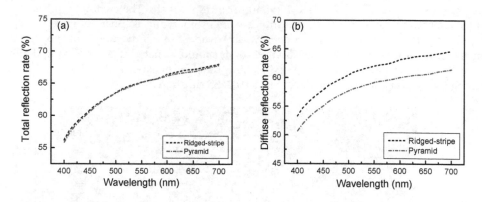

Fig. 17. The TR and DR rates versus the wavelength curves for ridged-stripe and pyramid textured 304BA SS substrates.

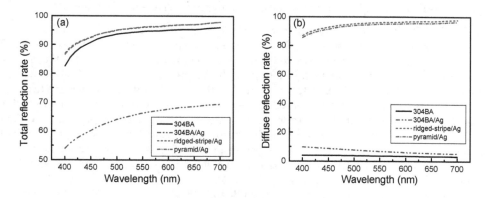

Fig. 18. The TR and DR rates versus the wavelength curves for Ag films coated/untreated 304BA SS substrate and Ag film coated/ridged-stripe and pyramid textured 304BA SS substrates.

Fig. 19 shows the relationship between the DR/TR rates and the total effective area of the Ag film coated/textured 304BA SS substrate. It should be noted that the total effective area was defined by the incident light reaching the textured 304BA SS substrate in an area of 100×100 µm². For example, the total effective area of the stripe textured 304BA SS substrate was calculated by the etched side wall area added to the untreated area of 10000 µm². For the ridged-stripe textured 304BA SS, the total effective area was calculated by summing the nine ridged-surfaces within an area measuring 100×100 µm². For the pyramid textured 304BA SS substrate, the total effective area was calculated by adding 25 pyramid-textured surfaces to the no-pyramid-coverage areas. Since the high reflection property of Ag films, the TR rate was almost higher than 90% after Ag-film coating of the textured 304BA SS substrates. It is worth noting that the DR rate increased linearly with the increase in total effective area of the stripe-textured 304BA SS substrate. However, the increase of the DR rate with the increase in the total effective area for the ridged-stripe and pyramid textured 304BA SS substrate was much more dramatic. We believe that the dramatic increase in the DR rate was due to the fact that the textured surface generated a random distribution of light by reflection from the textured surface. The aspect ratio for the ridged-stripe and pyramidal textured 304BA SS substrate were about 1/3.5 with an opening angle of 120°. In addition, the diffuse rate was defined when the incident light angle was zero, and the reflection light of that angle was larger than 8° over the incident light. Thus, the increased light diffuse due to the 120° opening angle of the texture surface caused the dramatic increase of the DR rate for the ridged-stripe and pyramid textured 304BA SS substrate. In addition, weakly absorbed light is totally reflected internally at the top surface of the cell as long as the angle of incidence inside the a-Si at the a-Si/TCO interface is greater than 16° (Banerjee and Guha 1991). It was indicated that the tilt angle of the V-shaped light trapping configuration substantially increases the photocurrent generation efficiency (Rim et al. 2007). The photocurrent increased with the increase of the tilt angle of the V-shaped configuration and is believed to enhance the number of ray bounces per unit cell area over that in a planar structure at each point in the V-fold structure. Therefore, the tilted angle of the textured surface is related to the DR and TR rate, and must be carefully investigated in future study.

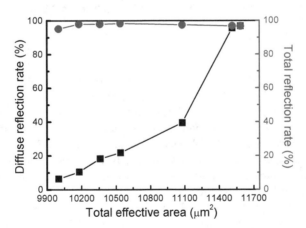

Fig. 19. The TR and DR rates as a function of the total effective area for Ag films coated on textured 304BA SS substrates.

4. Conclusions

We have demonstrated that a large diameter or a small interval of a concave shaped structure made from textured 430BA SS substrate can improve the DR rate of light.

However, the textured surface of a 430BA SS substrate led to a lower TR rate compared to a specular surface of raw 430BA SS substrate. This was due to the trapping of light in the hollows of the highly textured surface. Moreover, coating the textured 430BA SS substrate with an Ag film substantially improved not only the DR rate but also the TR rate of the incident light. The slow increase of the TR and DR rates versus the wavelength in the IR region of the Ag coated/textured 430BA SS substrates was due to the Ag absorption effect. We believe that Ag coated/textured 430BA SS substrates can generate a random distribution of light, increase the light trapping efficiency and be applied in thin films solar cells.

In addition, the DR and TR rate of the stripe, ridged-stripe and pyramid textured 304BA SS substrate were investigated to determine the optimal surface for increasing their light trapping efficiency. The DR rate increased with the increase in the total effective area of the Ag film coated/stripe textured 304BA SS substrate. It is believed that the tilt angle of the textured 304BA SS substrate increases the DR rate. The experimental results showed that the DR rate and the TR rate of the Ag film coated/ ridged-stripe textured 304BA SS substrate can achieve up to ~97% and 98% efficiency, respectively. The DR and TR rate of the Ag film coated/ridged-stripe textured 304BA SS substrates increased 28-fold and 1.4-fold, respectively, compared with the untreated 304BA SS substrate. The drastically increased DR rate is due to not only the increase in total effective area, but also to the decrease in the opening angle of the ridged textured substrate which generates a more random distribution of light by scattering.

5. Acknowledgment

The authors gratefully acknowledge the financial support from the National Science Council of Taiwan, R.O.C. under Contract No. NSC-98-2112-M155-001-MY3 and NSC-99-2221-E-155-065.

6. References

Banerjee A. and S. Guha. (1991). Study of back reflectors for amorphous silicon alloy solar cell application. *J. Appl. Phys.*, Vol. 69, pp. 1030., ISSN: 1089-7550

Curtin Benjamin, Rana Biswas, and Vikram Dalal. (2009). Photonic crystal based back reflectors for light management and enhanced absorption in amorphous silicon solar cells. *Appl. Phys. Lett.* Vol. 95, pp. 231102., ISSN: 1077-3118

Chau Joseph Lik Hang, Ruei-Tang Chen, Gan-Lin Hwang, Ping-Yuan Tsai and Chien-Chu Lin. (2010). Transparent solar cell window module. *Sol. Energy Mater. Sol. Cells.*, Vol. 94, pp. 588., ISSN: 0927-0248.

Deckman H. W., C. R. Wronski, H. Wittzke, and E. Yablonovitch. (1983). Optically enhanced amorphous silicon solar cells. *Appl. Phys. Lett.*, Vol. 42, pp. 968., ISSN: 1077-3118.

Ferlanto A. S., G. M. Ferreira, J. M. Pearce, C. R. Wronski, R. W. Collins, X. Deng, and G. Ganguly. (2002). Analytical model for the optical functions of amorphous semiconductors from the near-infrared to ultraviolet: Applications in thin film photovoltaics. *J. Appl. Phys.*, Vol. 92, pp. 2424., ISSN: 1089-7550

Fung Taddy Y. Y. and H. Yang. (2008). Study on thermal performance of semi-transparent building-integrated photovoltaic glazings. *Energy and Buildings.*, Vol. 40, pp. 341-350., ISSN: 0378-7788.

Hartstein A., J. R. Kirtley, J. C. Tsang. (1980). Enhancement of the Infrared Absorption from Molecular Monolayers with Thin Metal Overlayers. *Phys. Rev. Lett.*, Vol. 45, pp. 201. , ISSN: 1079-7114

Hatta A., T. Ohshima, W. Suëtaka. (1982). Observation of the enhanced infrared absorption of p-nitrobenzoate on Ag island films with an ATR technique. *J. Appl. Phys. A.*, Vol. 29, pp. 71., ISSN: 2158-3226

Huang B. R., K. H. Chen, W. Z. Ke. (2000). Surface-enhanced Raman analysis of diamond films using different metals. *Mater. Lett.*, Vol.42, pp. 162., ISSN:0167-577X

He Chun, Ya Xiong, Jian Chen, Changhong, Xihai Zhu. (2003). Photoelectrochemical performance of Ag–TiO$_2$/ITO film and photoelectrocatalytic activity towards the oxidation of organic pollutants. *J. Photochem. Photobiol. A.*, Vol. 57, pp. 71., ISSN: 1010-6030

Jenkins F. A. and H. E. White. (McGraw Hill, New York, 1957). *Fundamentals of Optics.*, P. 522. ISBN: 0070-8534-60

John A. E. St., U. S. Patent No. 3 487 223

Llopis F. and I. Tobías. (2005). The role of rear surface in thin silicon solar cells. *Sol. Energy Mater. Sol. Cells.*, Vol. 87, pp. 481., ISSN: 0927-0248

Lee Shuo Jen, Shiow Long Chen, Cheng Wei Peng, Chih Yuan Lin, Wen Cheng Ke. (2009). Enhanced diffuse reflection of light into the air using silver coating on periodically textured 430BA stainless steel substrate. *Mater. Chem. Phys.*, Vol. 118, pp. 219-222., ISSN: 0254-0584

Müller J., B. Rech, J. Springer and M. Vanecek (2004). TCO and light trapping in silicon thin film solar cells. *Sol. Energy.* Vol. 77, pp. 917., ISSN: 0038-092X

Nishikawa Y., T. Nagasawa, K. Fujiwara, M. Osawa. (1993). Silver island films for surface-enhanced infrared absorption spectroscopy: effect of island morphology on the absorption enhancement. *Vib. Spectrosc.*, Vol. 6, pp. 43., ISSN: 0924-2031

Otte K., L. Makhova, A. Braun, I. Konovalov. (2006). Flexible Cu(In,Ga)Se$_2$ thin-film solar cells for space application. *Thin Solid Films.*, Vol. 511, pp. 613., ISSN: 0040-6090

Osawa M. (1997). Dynamic Processes in Electrochemical Reactions Studied by Surface-Enhanced Infrared Absorption Spectroscopy (SEIRAS). *Bull. Chem. Soc. Jpn.*, Vol. 70, pp. 2861., ISSN: 0009-2673

Rech B., O. Kluth, T. Repmann, T. Roschek, J. Springer, J. Müller, F. Finger, H. Stiebig, and H. Wagner. (2002). New materials and deposition techniques for highly efficient silicon thin film solar cells. *Sol. Energy Mater. Sol. Cells.*, Vol. 74, pp. 439., ISSN: 0927-0248

Rim Seung-Bum, Shanbin Zhao, Shawn R. Scully, Michael D. McGehee and Peter Peumans. (2007). An effective light trapping configuration for thin-film solar cells. *Appl. Phys. Lett.* Vol. 91, pp. 243501. ISSN: 1077-3118

Selvan J. A. Anna., A. E. Delahoy, S. Guo and Y. M. Li. (2006). A new light trapping TCO for nc-Si:H solar cells. *Sol. Energy Mater. Sol. Cells.*, Vol. 90, pp. 3371., ISSN: 0927-0248

Sőderstrőm T., F. -J. Haug, V. Terrazzoni-Daudrix, and C. Ballif, J. (2008). Optimization of amorphous silicon thin film solar cells for flexible photovoltaics. *J. Appl. Phys.*, Vol. 103, pp. 114509-1., ISSN: 1089-7550

Yablonovitch E. and G. Cody. (1982). Intensity enhancement in textured optical sheets for solar cells. *IEEE Trans. Electron. Devices ED.*, Vol. 29, pp. 300., ISSN: 0018-9383

Zhou Dayu and Rana Biswas. (2008). Photonic crystal enhanced light-trapping in thin film solar cells. *J. Appl. Phys.*, Vol. 103, pp. 093102. , ISSN: 1089-7550

4

Electrodeposited Cu$_2$O Thin Films for Fabrication of CuO/Cu$_2$O Heterojunction

Ruwan Palitha Wijesundera
Department of Physics, University of Kelaniya, Kelaniya
Sri Lanka

1. Introduction

Solar energy is considered as the most promising alternative energy source to replace environmentally distractive fossil fuel. However, it is a challenging task to develop solar energy converting devices using low cost techniques and environmentally friendly materials. Environmentally friendly cuprous oxide (Cu$_2$O) is being studied as a possible candidate for photovoltaic applications because of highly acceptable electrical and optical properties. Cu$_2$O has a direct band gap of 2 eV (Rakhshani, 1986; Siripala et al., 1996), which lies in the acceptable range of window material for photovoltaic applications. It is a stoichiometry defect type semiconductor having a cubic crystal structure with lattice constant of 4.27 Å (Ghijsen et al., 1988; Wijesundera et al., 2006). The theoretical conversion efficiency limit for Cu$_2$O based solar cells is about 20% [5].

Thermal oxidation was a most widely used method for the preparation of Cu$_2$O in the early stage. It gives a low resistive, p-type polycrystalline material with large grains for photovoltaic applications. It was found that Cu$_2$O grown at high temperature has high leakage-current due to the shorting paths created during the formation of the material, and it causes low conversion efficiencies. Therefore it was focused to prepare Cu$_2$O at low temperature, which may provide better characteristics in this regard. Among the various Cu$_2$O deposition techniques (Olsen et al., 1981; Aveline & Bonilla, 1981; Fortin & Masson, 1981; Roos et al., 1983; Sears & Fortin, 1984; Rakhshani, 1986; Rai, 1988; Santra et al., 1992; Musa et al., 1998; Maruyama, 1998; Ivill et al., 2003; Hames & San, 2004; Ogwa et al., 2005), electrodeposition (Siripala & Jayakody, 1986, Siripala et al., 1996; Rakhshani & Varghese, 1987a, 1988b; Mahalingam et al., 2004; Tang et al., 2005; Wijesundera et al., 2006) is an attractive one because of its simplicity, low cost and low-temperature process and on the other hand the composition of the material can be easily adjusted leading to changes in physical properties. Most of the techniques produce p-type conducting thin films. Many theoretical and experimental studies (Guy, 1972; Pollack & Trivich, 1975; Kaufman & Hawkins, 1984; Harukawa et al., 2000; Wright & Nelson, 2002; Paul et al., 2006) have been revealed that the Cu vacancies originate the p-type conductivity. However, electrodeposition (Siripala & Jayakody, 1986, Siripala et al., 1996; Wijesundera et al., 2000; Wijesundera et al., 2006) of Cu$_2$O thin films in a slightly acidic aqueous baths produce n-type conductivity. Further it has been reported that the origin of this n-type behavior is due to oxygen vacancies and/or additional copper atoms. Recently, Garutara *et al.* (2006) carried out the photoluminescence (PL) characterisation for the electrodeposited n-type polycrystalline Cu_2O, and confirmed that the n-type conductivity is due to the oxygen vacancies created in the lattice. This n-type conductivity of Cu_2O is very important in developing low cost thin film solar cells because the electron affinity of Cu_2O is comparatively high. This will enable to explore the possibility of making heterojunction with suitable low band gap p-type semiconductors for application in low cost solar cells.

Most of the properties of the electrodeposited Cu_2O were reported to be similar to those of the thermally grown film (Rai, 1988). The electrodeposition of Cu_2O is carried out potentiostatically or galvanostatically (Rakhshani & Varghese, 1987a, 1988b; Mahalingam et al., 2000; Mahalingam et al., 2002). Dependency of parameters (concentrations, pH, temperature of the bath, deposition potential with deposits) had been investigated by several research groups (Zhou & Switzer, 1998; Mahalingam et al., 2002; Tang et al., 2005; Wijesundera et al., 2006). The results showed that electrodeposition is very good tool to manipulate the deposits (structure, properties, grain shape and size, etc) by changing the parameters. Various electrolytes such as cupric sulphate + ethylene glycol alkaline solution, cupric sulphate aqueous solution, cupric sulphate + lactic acid alkaline aqueous solution, cupric nitrate aqueous solution and sodium acetate+ cupric acetate aqueous solution, have been reported in the electrodeposition of Cu_2O.

Cu_2O-based heterojunctions of ZnO/Cu_2O (Herion et al., 1980; Akimoto et al., 2006), CdO/Cu_2O (Papadimitriou et al., 1981; Hames & San, 2004), ITO/Cu_2O (Sears et al., 1983), TCO/Cu_2O (Tanaka et al., 2004), and Cu_2O/Cu_xS (Wijesundera et al., 2000) were studied in the literature, and the reported best values of V_{oc} and J_{sc} were 300 mV and 2.0 mA cm^{-2}, 400 mV and 2.0 mA cm^{-2}, 270 mV and 2.18 mA cm^{-2}, 400 mV and 7.1 mA cm^{-2}, and 240 mV and 1.6 mA cm^{-2}, respectively.

Cupric oxide (CuO) is one of promising materials as an absorber layer for Cu_2O based solar cells because it is a direct band gap of about 1.2 eV (Rakhshani, 1986) which is well matched as an absorber for photovoltaic applications. It is also stoichiometry defect type semiconductor having a monoclinic crystal structure with lattice constants a of 4.6837 Å, b of 3.4226 Å, c of 5.1288 Å and β of 99.54° (Ghijsen et al., 1988). CuO had been wildly used for the photocatalysis applications. However, CuO as photovoltaic applications are very limited in the literature. The photoactive CuO based dye-sensitised photovoltaic device was recently reported by the Anandan et al. (2005) and we reported the possibility of fabricating the p-CuO/n-Cu_2O heterojunction (Wijesundera, 2010).

2. Growth and characterisation of electrodeposited Cu_2O

Electrodeposition is a simple technique to deposit Cu_2O on the large area conducting substrate in a very low cost. Electrodeposition of Cu_2O from an alkaline bath was first developed by Starek in 1937 (Stareck, 1937) and electrical and optical properties of electrodeposited Cu_2O were studied by Economon (Rakhshani, 1986). Rakshani and co-workers studied the electrodeposition process under the galvanostatic and potentiostatic conditions using aqueous alkaline $CuSO_4$ solution, to investigate the deposition parameters and properties of the material. Properties of the electrodeposited Cu_2O were reported to be similar to those of the thermally grown films (Rai, 1988) except high resistivity. Siripala et al. (Siripala & Jayakody, 1986) reported, for the first time, the observation of n-type photoconductivity in the Cu_2O film electrodes prepared by the electrodeposition on various metal substrates in slightly basic aqueous $CuSO_4$ solution in 1986. However, we have reported that electrodeposited Cu_2O thin films in a slightly acidic acetate bath attributed n-type conductivity.

Potentiostatic electrodeposition of Cu_2O thin films on Ti substrates can be investigated using a three electrode electrochemical cell containing an aqueous solution of sodium acetate and cupric acetate. Cupric acetate are used as Cu^{2+} source while sodium acetate are added to the solution making complexes releasing copper ions slowly into the medium allowing a uniform growth of Cu_2O thin films. The counter electrode is a platinum plate and reference electrode is saturated calomel electrode (SCE). Growth parameters (ionic concentrations, temperature, pH of the bath, and deposition potential domain) involved in the potentiostatic electrodeposition of the Cu_2O thin films can be determined by the met-hod of voltommograms.

voltammetric curves were obtained in a solution containing 0.1 M sodium acetate with the various cupric acetate concentrations, while temperature, pH and stirring speed of the baths were maintained at values of 55 ºC, 6.6 (normal pH of the bath) and 300 rev./min respectively. Curve a) in Fig. 1 is without cupric acetate and curves b), c) and d) are cupric acetate concentrations of 0.25 mM, 1 mM and 10 mM respectively. Significant current increase can not be observed in absence with cupric acetate and cathodic peaks begin to form with the introduction of Cu^{2+} ions into the electrolyte. Two well defined cathodic peaks are resulted at –175 mV and –700 mV Vs SCE due to the presence of cupric ions in the electrolyte and these peaks shifted slightly to the anodic side at higher cupric acetate concentrations. First cathodic peak at –175 mV Vs SCE attributes to the formation of Cu_2O on the substrate according to the following reaction.

$$2Cu^{2+} + H_2O + 2e^- \Rightarrow Cu_2O + 2H^+$$

Second cathodic peak at –700 mV Vs SCE attributes to the formation of Cu on the substrate according to the following reaction.

$$Cu^{2+} + 2e^- \Rightarrow Cu$$

By examining the working electrode, it can be observed that the electrodeposition of deposits on the substrate is possible in the entire potential range. However, as revealed by the curves in Fig. 1, at higher concentrations the peaks are getting broader and therefore the formation of Cu and Cu_2O simultaneously is possible at intermediate potentials (curve d of Fig. 1). The deposition current slightly increases and the peaks are slightly shifted to the positive potential side as increasing the bath temperature range of 25 ˚C to 65 ˚C.

Fig. 2 shows the dependence of the voltammetric curves on the pH of the deposition bath. It is seen that cathodic peak corresponding to the Cu deposition is shifted anodically by about 500 mV and cathodic peak corresponding to the Cu_2O deposition is shifted anodically by about 100 mV. This clearly indicates that acidic bath condition favours the deposition of copper over the Cu_2O deposition and the possibility of simultaneous deposition of Cu and Cu_2O even at lower cathodic potentials. This is further investigated in the following sections.

The potential domain of the first cathodic peak gives the possible potentials for the electrodeposition of Cu_2O films while second cathodic peak evidence the possible potential domain for the electrodeposition of Cu films. It is evidence that Cu_2O can be electrodeposited in the range of 0 to -300 mV Vs SCE and Cu can be electrodeposited in the range of -700 to -900 mV Vs SCE. The potential domains of the electrodepostion of Cu_2O and Cu are independent of the Cu^{2+} ion concentration and the temperature of the bath. However, the deposition rate is increased with the increase in the concentration or the temperature of the bath.

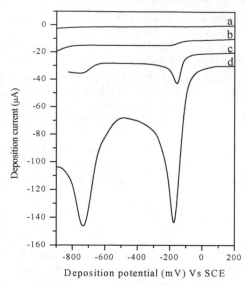

Fig. 1. Voltammetric curves of the Ti electrode (4 mm²) obtained in a solution containing
0.1 M sodium acetate and cupric acetate concentrations of a) 0 mM, b) 0.25 mM, c) 1 mM and
d) 10 mM

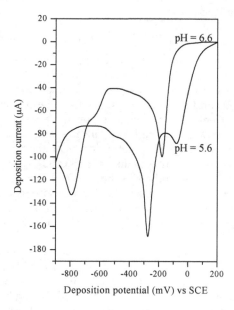

Fig. 2. Voltammetric curves of the Ti electrode (4 mm²) in an electrochemical cell containing
0.1 M sodium acetate and 0.01 M cupric acetate solutions at two different pH values
(pH was adjusted by adding diluted HCl).

Cu_2O film deposition potential domain can be further verified by the X-ray diffraction (XRD) spectra obtained for the films electrodeposited at various potentials (-100 to -900 mV Vs SCE). Fig. 3 shows the XRD spectra of the films deposited at a) -200 mV Vs SCE, b) -600 mV Vs SCE and c) -800 mV Vs SCE on Ti substrates in a bath containing 0.1 M sodium acetate and 0.01 M cupric acetate aqueous solution. Fig. 3(a) shows five peaks at 2θ values of 29.58°, 36.43°, 42.32°, 61.39° and 73.54° corresponding to the reflections from (110), (111), (200), (220) and (311) atomic plans of Cu_2O in addition to the Ti peaks. Fig. 3(b) exhibits three additional peaks at 2θ values of 43.40°, 50.55° and 74.28° corresponding to the reflection from (111), (200) and (220) atomic plans of Cu in addition to the peaks corresponding to the Cu_2O and Ti substrate. It is evident that the intensity of Cu peaks increases with increase of the deposition potential with respect to the SCE while decreasing the intensities of Cu_2O peaks. Peaks corresponding to the Cu_2O disappeared with further increase in deposition potential. XRD of Fig. 3(d) exhibits peaks corresponding to Cu and Ti only. Thus, in the acetate bath single phase polycrystalline Cu_2O thin films with a cubic structure having lattice constant 4.27 Å are possible only with narrow potential domain of 0 to -300 mV Vs SCE while Cu thin films having lattice constant 3.61 Å are possible at potential –700 mV and above Vs SCE.

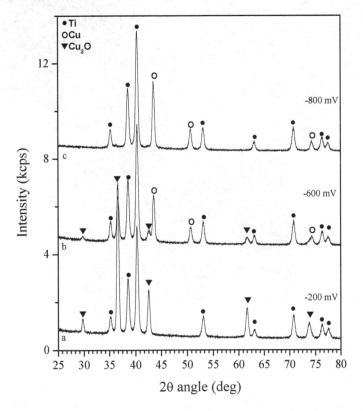

Fig. 3. XRD spectra obtained for the films deposited on Ti substrate at the potentials (a) -200 mV Vs SCE, (b) -600 mV Vs SCE and (c) -800 mV Vs SCE

Fig. 4 shows the scanning electron micrographs (SEMs) of the above set of samples. It is evident that the surface morphology depends on the deposition potential and the films grown on Ti substrate are uniform and polycrystalline. Grain size of Cu_2O is in the range of ~1-2 μm. It is observed that the Cu_2O thin film deposited at –200 mV Vs SCE exhibit cubic structure (Fig. 4(a)) and deviation from the cubic structure can be observed when deposition potential deviate from the -200 mV Vs SCE. Thus, polycrystalline Cu_2O thin films with cubic grains are possible only within a very narrow potential domain of around –200 mV Vs SCE. Fig. 4(b) shows the existence of spherical shaped Cu on top of Cu_2O when film deposited at –400 mV Vs SCE. The co-deposition of Cu with Cu_2O is evident in the XRD spectra, too. This small grains of Cu distributed over the Cu_2O surface will be useful in some other applications. It is clear from XRD and SEM results that Cu_2O, Cu_2O + Cu, and Cu microcrystalline thin films can be separately electrodeposited on Ti substrate by changing the deposition potential from –100 mV to –900 mV Vs SCE using the same electrolyte.

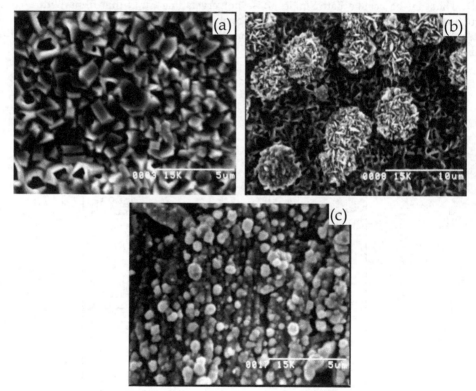

Fig. 4. Scanning electron micrographs of thin films electrodeposited at (a) -200 mV Vs SCE, (b) -400 mV Vs SCE and (c) -800 mV Vs SCE

Cu_2O thin films produce negative photovoltages in a photelectrochemical cell (PEC) containing 0.1 M sodium acetate under the white light illumination of 90 W/m². Active area of the film in a PEC was ~1 mm². The magnitudes of the photovoltage and the photocurrent of Cu_2O films deposited at –100 mV to –500 mV Vs SCE were 125 mV and 5 μA, 168 mV and 6.5 μA, 172 mV and 8 μA, 210 mV and 15 μA and 68 mV and 1 μA respectively. Also Cu_2O

film deposited at –600 mV Vs SCE shows the photoactivity but magnitudes of the photovoltage and photocurrent were very small. The best photoresponse we have obtained for the Cu$_2$O thin film deposited at –400 mV Vs SCE. This may be due to the better charge transfer process between Cu$_2$O and electrolyte due to the randomly distributed Cu spheres on top of Cu$_2$O thin films as shown in Fig. 4.

The optical absorption measurements of the Cu$_2$O thin films on indium doped tin oxide (ITO) substrate deposited at -100 mV to -600 mV Vs SCE indicate that the electrodeposited Cu$_2$O has a direct band gap of 2.0 eV, and the band gap of the material is independent of the deposition potential.

Photoactivity of the films was further studied by the dark and light current-voltage measurements. Fig. 5 shows the dark and light current-voltage characteristics in a PEC of the films deposited at (a) –200 mV and (b) –400 mV Vs SCE. Current-voltage measurements were obtained in three electrode electrochemical cell. The change of the sign of the photocurrent with the applied voltage shows the evidence for the existence of two junctions within the Ti/Cu$_2$O/electrolyte system. Particularly with the positive applied bias voltage, the Cu$_2$O/electrolyte junction become dominant and thereby the n-type photosignal is produced, when negative bias voltage is applied the Ti/Cu$_2$O junction become dominant and therefore a p-type signal is produced. Similar results have been reported earlier on the ITO/Cu$_2$O/electrolyte system (Siripala et al., 1996) and ITO/Cu$_2$O/Cu$_x$S system (Wijesundera et al., 2000). It has been reported earlier that both n- and p-type photosignals can be obtained in the currant–voltage scans due to the existence of Ti/Cu$_2$O and Cu$_2$O/electrolyte Schottky type junctions. The enhancement of n-type signal could be due to the enhancement of Cu$_2$O/electrolyte junction as compared with the Ti/Cu$_2$O junction.

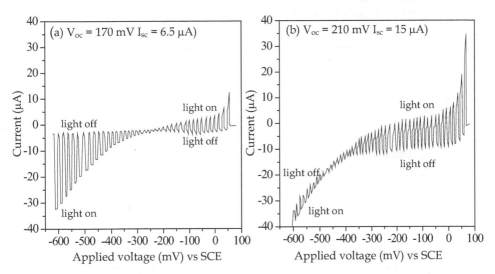

Fig. 5. Dark and light current-voltage characteristics for the films deposited at (a) -200 mV and (b) -400 mV Vs SCE in a PEC containing 0.1 M sodium acetate under the white light illumination of 90 W/m² (effective area of the film is ~1 mm²).

Single phase polycrystalline n-type Cu$_2$O thin films can be potentiostatically electrodeposited on conducting substrates selecting proper deposition parameters and these films are uniform and well adhered to substrate. Garutara *et al.* (Garuthara & Siripala, 2006) carried out the photoluminescence (PL) characterisation for the electrodeposited n-type polycrystalline Cu_2O. They showed the existence of the donor energy level of 0.38 eV below the bottom of the conduction band due to the oxygen vacancies and confirmed that the n-type conductivity is due to the oxygen vacancies created in the lattice. Previously reported electrodeposited Cu_2O in a various deposition bath, except slightly acidic acetate bath, attribute p-type conductivity due to the Cu vacancies created in the lattice as thermally grown films.

3. Growth and characterisation of CuO thin films

It is expected that Cu_2O thin films can be oxidized by the annealing in air and thus converted into CuO. Therefore, annealing effects of the electrodeposited Cu_2O thin films in air were investigated in order to obtain a single phase CuO thin films on Ti substrate. Cu_2O thin films on Ti substrates were prepared under the potentiostatic condition of -200 mV Vs SCE for 60 min. in the three electrode electrochemical cell containing 0.1 M sodium acetate and 0.01 M cupric acetate aqueous solution. Temperature of the bath was maintained to 55 °C and the electrolyte was continuously stirred using a magnetic stirrer. All the thin films are uniform and having a thickness of about 1 μm which was calculated by monitoring the total charge passed during the film deposition through the working electrode (WE).

The bulk structure of the films, which were annealed at different temparatures and durations, can be determined by XRD measurements and Fig. 6 shows the XRD spectra of the films annealed at 150 to 500 °C in air, in addition to the as grown Cu_2O. Results show that Cu_2O structure remains stable even though films are annealed at 300 °C, as reported by Siripala *et al.* (1996). Formation of CuO structure can be observed when films are aneealed at

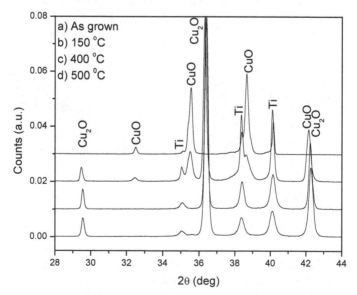

Fig. 6. X-ray diffraction patterns of electrodeposited Cu_2O thin films a) as grown and annealed at b) 150 ᵒC, c) 400 ᵒC and d) 500 ᵒC

400 ºC for 15 min. Fig. 6 shows that the intensities of the peaks correspondent to the CuO structure increases while intensities of the peaks correspondent to the Cu_2O structure decreases with the increasing of annealing temperature and duration. The reflections from the Cu_2O structure disappear when the film is annealed at 500 ˚C for 30 min. in air. It is reveled that the single phase CuO thin films on Ti substrate can be prepared by annealing Cu_2O in air.

The surface morphology of the annealing Cu_2O thin films is studied with SEMs. Fig. 7 shows SEMs of (a) as grown, and annealed in air at (b) 175 ˚C, (c) 400 ˚C and (d) 500 ˚C. Results reveal that, by increasing the annealing temperature, the size of the cubic shape polycrystalline grain gradually increase up to 200 ˚C, change to the different shape at 400 ºC and converted to the monoclinic like shape polycrystalline grain at 500 ºC. Cu_2O thin films have the cubic-like polycrystalline grains. SEMs clearly show that structural phase transition take place from Cu_2O, Cu_2O-CuO, CuO as reveal by the XRD patterns. CuO crystallites are in the order of 250 nm.

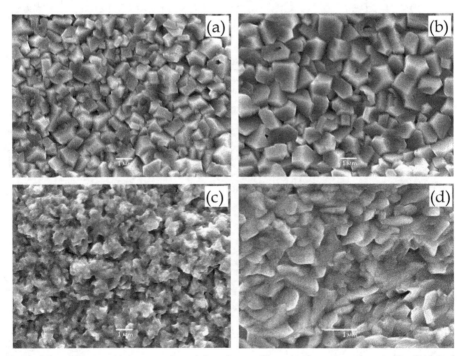

Fig. 7. Scanning electron micrographs of the electrodeposited semiconductor Cu_2O thin films a) as grown and annealed in air at (b) 175 ˚C, (c) 400 ˚C and (d) 500 ˚C

Photosensitivity (V_{oc} and I_{sc}) of the annealed electrodeposited Cu_2O thin films in a two electrode PEC cell containing 0.1 M sodium acetate aqueous solution, under white light illumination of 90 W/m², shows that initial n-type photoconductivity changes to the p-type after annealing 300 ºC. Type of the photoconductivity of the Cu_2O thin films can be converted from n- to p-type with annealing because of Cu_2O structure remain same even if films annealed at 300 ºC as revealed by XRD patterns.

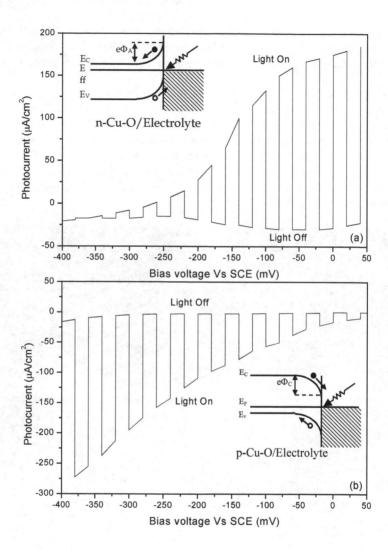

Fig. 8. Dark and light current voltage characteristics of electrodeposited Cu_2O thin film electrodes annealed at (a) 250 °C and (b) 300 °C. Energy level diagrams for n-type and p-type Cu_2O films in the electrolyte are shown in the insets, where the electron, hole, anodic and cathodic potential are denoted by solid and open circles, $e\Phi_A$ and $e\Phi_C$ respectively.

The photoactivity of the thin films has been further studied by the dark and light current voltage characteristics in a three electrode electrochemical cell. The counter and the reference electrodes are Pt plate and SCE, respectively. The bias voltage has been applied to the working electrode (Ti/Cu_2O) with respect to the SCE. Fig. 8 shows the dark and light current-voltage characteristics of the thin films annealed at (a) 250 °C and (b) 300 °C. The similar behaviour is observed for the thin films annealed at less than 250 °C and annealed at grater than 300 °C, respectively, and is reproducible for each film. In Fig. 8(a), the anodic photocurrent increases with increasing the anodic potential. This suggests that the n-type photoconductivity is due to an anodic potential behaviour, and is reproducibly observed for the thin films annealed at < 250 °C. This suggests that the n-type photoconductivity is due to the anodic potential barrier formed at the semiconductor/electrolyte interface, as the inset of Fig. 8(a). However, the photocurrent-potential behaviour is completely changed for the film annealed at ≥ 300 °C. In Fig. 8(b), the cathodic photocurrent results from the cathodic potential barrier formed at the interface, as shown in the inset. This cathodic photoresponse assures that the electrical conductivity of the electrodeposited Cu_2O films can be changed from the n-type to p-type property by annealing in air. Fig. 9 shows the dark and light current-voltage characteristics of the CuO thin film in a PEC cell containing 0.1 M sodium acetate aqueous solution. The cathodic photocurrent is produced in the range from the anodic to cathodic bias potentials, and the cathodic photocurrent increases with increasing the cathodic potential. This suggests that the p-type photoconductivity is due to the cathodic potential barrier forms at the semiconductor/electrolyte interface. It reveals that the electrodeposited CuO thin films are p-type semiconductors.

Structural phase transition from Cu_2O to CuO with annealing and the quality of thin films can be further investigated using Extended X-ray Absorption Fine Structure (EXAFS) which gives local structure around Cu ions. Fig. 10 shows the X-ray absorption spectra (XAS) in the region of 8800 to 9430 eV near the Cu-K edge for the thin films, annealed at 150, 400, and 500 °C by using the florescence detection (FD) method. XAS suggest that the local structures around Cu ions in the annealed Cu_2O thin films are remain same when films annealed at less than 300 °C and significantly different when films annealed at grater than 300 °C.

Refinements of a Fourier transformation spectrum $|F(R)|$ obtained from the oscillating EXAFS spectra can be used to study the quality of Cu_2O and CuO thin films. Fig. 11, solid circles show the observed $|F(R)|$ of the thin film annealed at 150 °C, where the abscissa is a radial distance (R(Å)) from a X-ray absorbing Cu ion to its surrounding cations and anions. Fig. 11, a solid line shows a theoretical $|F(R)|$. The refinement produces a good fit between the observed and theoretical $|F(R)|$ indicating the local structure around Cu ions of the film is verymuch similar to the ideal Cu_2O structure. Fig. 12 is similar refinement for CuO thin film. These results convince that the thin films are high quality single phase Cu_2O and CuO structures (free of amorphous phases and impurities). Detail investigation has been reported (Wijesundera et al., 2007).

It is characterised that single phase Cu_2O thin films are converted to two phase Cu_2O and CuO composit films with increasing the annealing temperature. Single phase CuO thin films can be obtained by annealing at 500 °C for 30 min in air. Extended X-ray absorption fine structure (EXAFS) near the Cu K edge of the Cu_2O thin films (annealed at 150 °C for 15 min.) and CuO thin films (annealed at 500 °C for 30 min.) are confirmed that the films are high quality single phase Cu_2O and CuO (free of amophous phases) respectively. Conductivity type of the films strongly depends on the annealing treatment. n-type conductivity of the Cu_2O thin films are changed to p-type when the films are annealed at 300 °C. CuO thin films are photoactive and p-type in a PEC containing 0.1 M sodium acetate.

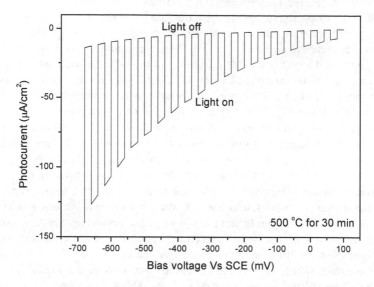

Fig. 9. Dark and light current voltage characterisation of CuO thin film in a PEC cell containing 0.1 M sodium acetate aqueous solution.

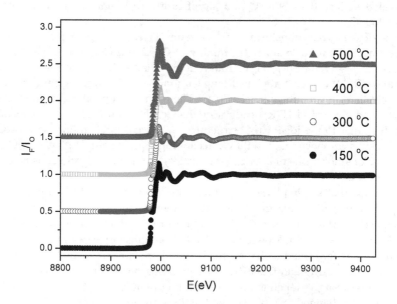

Fig. 10. X–ray absorption spectra of annealed Cu_2O thin at 150, 300, 400 and 500°C in air

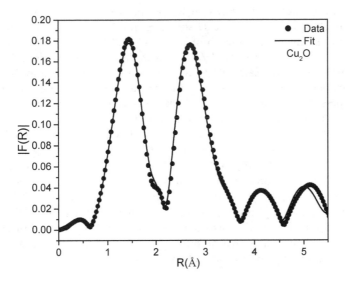

Fig. 11. Theoretical | F(R) | of the EXAFS spectrum at Cu K-edge obtained by the least squares refinement compared to the observed | F(R) | for the Cu$_2$O thin film annealed at 150°C

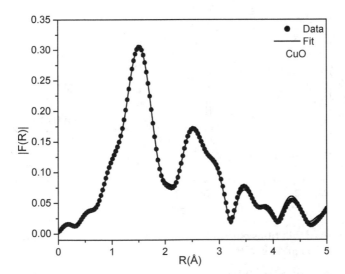

Fig. 12. Theoretical | F(R) | of the EXAFS spectrum at Cu K-edge obtained by the least squares refinement compared to the observed | F(R) | for the Cu$_2$O thin film annealed at 500°C

4. Fabrication and characterisation of CuO/Cu₂O heterojunction

In order to fabricate CuO/Cu$_2$O thin film hetorojunction, thin films of n-type Cu$_2$O are potentiostatically electrodeposited on a Ti substrate in an acetate bath and are annealed at 500 ºC for 30 min. in air for the growth of p-type CuO thin films. Thin films of Cu$_2$O are potentiostatically electrodeposited on Ti/CuO electrodes at different deposition potentials Vs SCE while maintaining the same electrolytic conditions, which used to deposit Cu$_2$O on the Ti substrate. Deposition period is varied form 240 min to 120 min in order to obtain sufficient thickness of the films. Film thickness was calculated by monitoring the total charge passed during the film deposition and it was ~1 μm.

Bulk structures of the electrodeposited films on Ti/CuO were studied by the XRD patterns. Fig. 13 shows the XRD patterns of the films deposited on Ti/CuO electrodes at the deposition potentials of -250, -400, -550 and -700 mV Vs SCE. XRD patterns evidence the formation of Cu$_2$O for all deposition potentials on Ti/CuO electrodes while Cu deposition starts in addition to the Cu$_2$O when the film deposited at -700 mV Vs SCE. Single phase Cu$_2$O are possible at the deposition potentials less than -700 mV Vs SCE. XRD patterns further show that peak intensities corresponding atomic reflection of Cu$_2$O increase with deposition potential. It indicates that amount of Cu$_2$O deposit is increased by increasing deposition potential. This is further studied by using SEM.

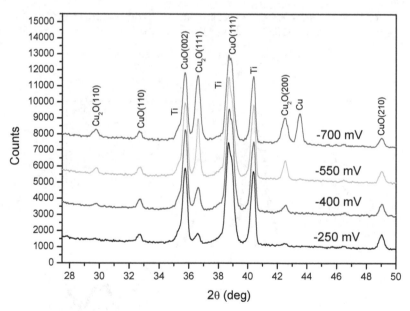

Fig. 13. XRD pattern of thin films electrodeposited on Ti/CuO electrode at the potentials -250, -400, -550 and -700 mV Vs SCE

The surface morphology of the films prepared on the Ti/CuO electrode at the different deposition potentials was studied using the SEMs in order to identify the Cu$_2$O thin film deposition conditions on Ti/CuO electrode. Figs. 14(a) to (c) show the SEMs of Cu$_2$O films deposited on the Ti/CuO at -250 to -550 mV Vs SCE. Fig. 14(a) shows the cubic shape Cu$_2$O grains on the CuO film and Figs. 14(a) to (c) show that the amount of Cu_2O increases with increasing the deposition potential. The SEMs reveal that the well covered Cu_2O layer can be deposited on Ti/CuO electrode under the potentiostatical condition of -550 mV Vs SCE and above. Grain size of the Cu_2O deposited on Ti substrate is in the range of ~ 1-2 µm as shown in Fig. 14(a) while it is lower to 1 µm when Cu_2O deposited on CuO at the deposition potential of -550 mV Vs SCE. The SEM with low magnification of Cu_2O deposited at -700 mV Vs SCE clearly shows the existence of Cu on the surface of Cu_2O as shown in the XRD pattern of the film deposited at -700 mV Vs SCE.

Fig. 14. Scanning electron micrograph of Cu_2O thin films electrodeposited on Ti/CuO electrode at (a) -250 mV, (b) -400 mV and (c) -550 mV Vs SCE

XRD and SEM reveal that well-covered single phase polycrystalline Cu_2O thin film on the Ti/CuO electrode can be possible at the deposition potential of −550 mV Vs SCE in an acetate bath. Structural matching of two semiconductors is very essential for fabricating a heterojunction. In general, the cubic-like Cu_2O grains and the monoclinic-like CuO grains are not match with each other to make the CuO/Cu_2O heterojunction. However, the electrodeposition technique produces the good matching of the Cu_2O grains to the monoclinic-like CuO grains. The shape of the grains can be easily changed when the electrodeposition technique is used to grow a semiconductor. The electrodeposition is a very good tool to fabricate the heterojunctions as it does not depend on the grain shape of the material. Further, the SEMs of the Cu_2O/CuO heterojunction suggested that the Cu_2O polycrystalline grains are grown from the surfaces of the CuO polycrystalline grains and make the good contacts between two thin film layers. For the completion of the device, very thin (few angstroms) Au grid consists of 1×8 mm^2 rectangular areas are deposited on Cu$_2$O using a vacuum sputtering technique. The electrical contacts to the Cu$_2$O surface (front contacts) is made using mechanically pressed transparent ITO plate to the Au grid, where the Ti substrate serves as back electrical contact to the CuO surface. The Ti/CuO/Cu$_2$O/Au heterojunction gave the open circuit voltage (V$_{oc}$) of 210 mV, short circuit current (J$_{sc}$) of 310 µA cm^2, fill factor (FF) of 0.26 and efficiency (η) of 0.02% under the white light illumination of 90 mWcm^{-2}. At the initial stage of fabrication the junction, the shape of the I–V characteristic as shown in Fig. 15 and values of V$_{oc}$ and J$_{sc}$ are encouraging despite the low photoactive performance of the heterojunction.

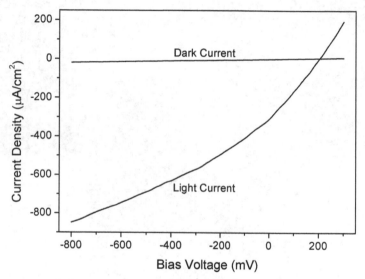

Fig. 15. Dark and light current-voltage characteristics of Ti/CuO/Cu$_2$O/Au heterojunction under the white light illumination of 90 mW/cm^2

For the better performance, very thin Cu$_2$O films should be used due to the high resistance of electrodeposited Cu$_2$O and should be find out better ohmic contact to the Cu$_2$O. Best omic contact to the n-type Cu$_2$O may be Al but not the Au. Au is very good omic contact to p-type Cu$_2$O. n-type electrodeposted Cu$_2$O has high resistivity is due to low doping density. Growth of the n-type Cu$_2$O with suitable dopent hasn't been achieved in the litterateur and it will be very important in developing Cu$_2$O based solar cells.

CuO/Cu$_2$O heterojunction was further investigated by means of X-ray diffractions and X-ray absorption spectra (XAS) at the Cu-K edge with grazing angle measurements. Layer by layer structural information of the CuO/Cu$_2$O heterojunction can be studied with grazing angle measurements. Fig. 16 shows the grazing angle (φ) dependency of the X-ray diffraction patterns of the CuO/Cu$_2$O heterojunction. The Ti peak of highest intensity at $2\theta = 40.23$ degree is indexed by (0,1,1) and (1,1,1) reflections of hexagonal structure and are not observed below $\varphi \sim 2.0$ degree. On the other hand, the reflections of Cu$_2$O and CuO structures are observed in all the grazing angles, though the reflections of Cu$_2$O structure shows the different grazing angle dependence to those of the CuO ones. Fig. 17 shows grazing angle dependency of (1,1,1) reflection of Cu$_2$O, (1,1,-1) reflection of CuO, (1,1,1) reflection of Ti, and intensity ratio of

$Cu_2O(1,1,1)$ and $CuO(1,1,-1)$ reflections. The results suggest that Cu_2O grain layer can only be observed below $\varphi = 0.1$ degree. With increasing the grazing angle, CuO grain layer can be observed gradually, as Ti-reflections. The bulk structural information of Cu_2O layer can be obtain for the grazing angles around 2.5 degree since it produces highest intensity of $(1,1,1)$ reflection of Cu_2O. It reveals that it is possible to obtain optimum structural information within the Cu_2O/CuO junction region in addition to the Cu_2O, CuO, and Ti for the grazing angles slightly grater than 2.5 degree. Further it is found that the intensity ratio of the $(1,1,1)$ reflection and $(1,1,-1)$ reflection is approximately constant above $\varphi = 5.0$ degree, in contrast to the intensity of the Ti-reflection. It shows that the bulk structural information of bi-layer (Cu_2O and CuO) can be studied at the grazing angle 5 degree or greater

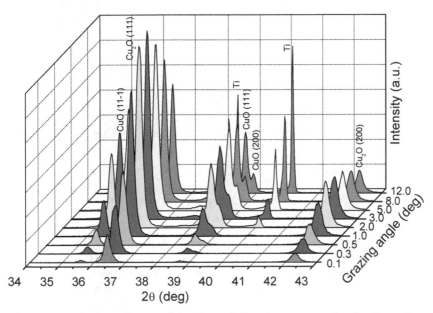

Fig. 16. Grazing angle dependency of the X-ray diffraction patterns for the electrodeposited Ti/CuO/Cu₂O heterojunction

The structural deformation localized around Cu ions of the CuO/Cu_2O heterojunction can be investigated from EXAFS. Fig. 18 shows the expanded partial XAS at Cu-K edge of $Ti/CuO/Cu_2O$ heterojunction at $\varphi = 0.3$ to 10.0 deg. The electrodeposited CuO/Cu_2O thin film heterojunction include Cu ions sited at different structures of Cu_2O and CuO. The spectra result from a convoluted XAS induced by interference between the X-ray photoelectron waves emitted by X-ray absorbing Cu ions and the backscattering waves of its surrounding ions for both structures. The grazing angle dependency of the XAS suggests that the incident X-ray beam penetrate the thin films of Cu_2O and CuO grains by the different path distance. It can be considered that the XAS measurements obtained at low grazing angles (0.3 and 0.5 deg) should be mainly the XAS of Cu_2O thin film, which is the front layer of the heterojunction. Therefore, the XAS at $\varphi = 0.5$ and 3.0 deg are compared with the observed XAS of the electrodeposited thin films of Cu_2O and CuO. Fig. 19 shows the comparsion of the expanded partial XAS at grazing angles of 0.5 and 3.0 deg and of the

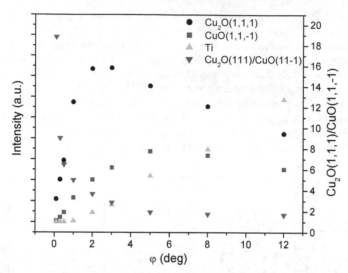

Fig. 17. Grazing angle dependency of the (1,1,1) reflection of the Cu$_2$O, (1,1,-1) reflection of the CuO structure, the (1,1,1) reflection of Ti and intensity ratio of Cu$_2$O (1,1,1) and CuO (1,1,-1) reflections

electrodeposited Cu$_2$O and CuO thin films, in addition to the calculated one of (0·5Cu$_2$O + 0·5CuO). However, XAS (local structure around Cu ions) at low grazing angles are not similar with Cu$_2$O. It shows that the XAS obtained at even low grazing angles are the convoluted spectra induced by the Cu$_2$O and CuO structures. The convolution effect of the XAS can be studied by fitting the observed XAS at φ = 0.5 deg from a simple mathematical convolution of Cu$_2$O-XAS and CuO-XAS. Fig. 19 shows that the observed XAS at φ = 0.5 deg is very similar to the calculated one of 0.5(Cu$_2$O-XAS) + 0.5(CuO-XAS). However, XAS at low grazing angles can be analyzed by a simple mathematical convolution of Cu$_2$O and CuO structures but not for the grazing angles higher than 0.5 deg. This reveals that the complex XAS results from the convoluted spectra induced from unknown structure in addition to the Cu$_2$O and CuO structures. The XAS modulation due to the unknown structure depends on the grazing angles, and the maximum XAS modulation appears at the grazing angle of 3.0 deg. This suggests that the junction region has very complex structure.

XAS of CuO/Cu$_2$O hetrojunction with different grazing angles can be further compared by studying corresponding Fourier transformations of the oscillating EXAFS spectra. Fig. 20 shows the observed |F(R)| for the bi-layer thin film of Ti/CuO/Cu$_2$O heterojunction at φ = 0·5 and 3·0 degrees and for the electrodeposited Cu$_2$O and CuO thin films with calculated one of (0·5Cu$_2$O + 0·5CuO). It is further confirmed that the |F(R)| obtained at φ = 0·5 and 3·0 degrees are not similar with that of Cu$_2$O structure and CuO one, but more complex. Comparison between the |F(R)| of the bi-layer thin film obtained at φ = 0·5 degree and the calculated one of (0·5Cu$_2$O + 0·5CuO) suggests that the |F(R)| of the bi-layer thin film is also convoluted by those of the Cu$_2$O and CuO structures. As in Fig. 20 peak amplitudes are very small for the |F(R)| at φ = 3.0 deg compared to the others. It implies that structure in the junction region is diluted one (the surrounding ions around the Cu absorbing ion do not well arranged). Results reveal that the formation of amorphous structure in the interface of CuO/Cu$_2$O heterojunction. It can be expected that amorphous

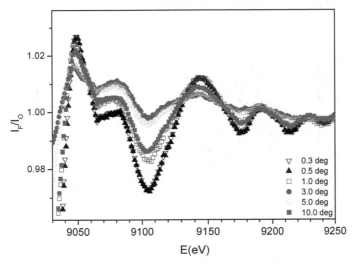

Fig. 18. Oscillating amplitudes I_F/I_o of the X-ray absorption spectra of the Ti/CuO/Cu₂O heterojunction at φ = 0·3 to 10·0 degrees

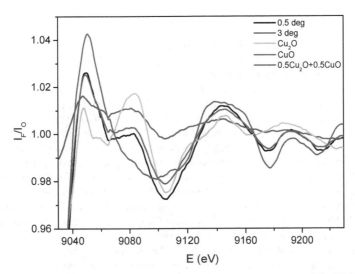

Fig. 19. Oscillating amplitudes I_F/I_o of the X-ray absorption spectra of the Ti/CuO/Cu₂O heterojunction at grazing angles of 0.5 and 3.0 degrees and of the electrodeposited Cu₂O and CuO thin films, in addition to the calculated one of (0·5Cu₂O + 0·5CuO)

structure formed in the middle of the CuO/Cu₂O heterojunction attributes better lattice matching between CuO and Cu₂O interface. Further, it can be considered that formation of the smooth energy band lineup at the interface of CuO/Cu₂O heterojunction without spikes at the conduction and valance bands (ΔE_c and ΔE_v). Band lineup between two semiconductors is a crucial parameter leading to better photoactive properties.

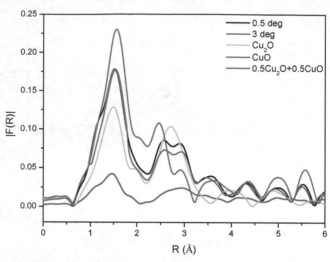

Fig. 20. Amplitudes | F(R) | obtained from the Fourier transformation of the EXAFS spectra of the Ti/CuO/Cu₂O heterojunction at φ = 0·5 and 3·0 degrees and of the electrodeposited Cu₂O and CuO thin films and the calculated one of (0·5Cu₂O + 0·5CuO)

5. Conclusion

Single phase polycrystalline Cu_2O thin films on Ti substrate can be electrodeposited using an acetate bath in a potential range of 0 to -300 mV Vs SCE. Thin films are well adherent to the Ti substrate and uniform having grain size of ~1-2 μm. Cu_2O deposited in an acetate bath at the pH of 6.6 produces n-type photoconductivity in a PV applications. The n-type photoconductivity of the as-grown Cu_2O thin film can be converted to p-type by annealing the films at 300 ºC in air. Therefore, it is reasonable to believe that the origin of the n-type Cu_2O is the oxygen ion vacancies created in the crystal lattice, and the conductivity-type conversion is due to the increment of the oxygen content in the lattice with annealing in air. Single phase polycrystalline CuO thin films can be prepared by annealing the electrodeposited Cu_2O at 500 ºC for 30 min in air. Films produce p-type photoresponse in a PEC. Well covered n-type Cu_2O thin film can be electrodeposied on Ti/CuO electrode at -550 mV Vs the SCE in an acetate bath. Films consist of microcrystallites of about 1 μm and are well adherent to the CuO. By depositing a suitable mettle grid on the Cu_2O thin film, p-CuO/n-Cu_2O heterojunction solar cell can be fabricated. The Ti/CuO/Cu₂O/Au heterojunction solar cell results in V_{oc} of 210 mV and J_{sc} of 310 μA. This initial stage performance can be enhanced by depositing very thin Cu_2O films leading to minimize the resistance of the Cu_2O and choosing better ohmic contact to the Cu_2O. Best ohmic contact to the n-type Cu_2O may be Al but not the Au. X-ray diffractions and the X-ray absorption spectra, using the synchrotron radiation, reveal that Cu_2O and CuO are high quality semiconducting thin films (free of amorphous phases) but amorphous structure is formed between CuO and Cu_2O while Cu_2O deposition on CuO.

6. Acknowledgment

Prof. W. Siripala, Professor of Physics, Department of Physics, University of Kelaniya, Sri Lanka and Prof. M. Hidaka, Department of Fundamental Physics, Graduate School of

Science, Kyushu University, Japan are gratefully acknowledged for their invaluable advice, guidance and encouragement.

7. References

Akimoto, K., Ishizuka, S., Yanagita, M., Nawa, Y., Goutam K. P. & Sakurai, T. (2006). Thin film deposition of Cu$_2$O and application for solar cells. *Sol. Energy*, Vol. 80, 715-722

Anandan, S., Wen, X. & Yang, S. (2005). Room temperature growth of CuO nanorod arrays on copper and their application as a cathode in dye-sensitized solar cells. *Mater. Chem. Phys.*, Vol. 93, 35-40

Aveline, A. & Bonilla, I. R. (1981). Spectrally selective surfaces of cuprous oxide (Cu$_2$O). *Sol. Energy Mater.*, Vol. 5, 2, 211-220

Fortin, E. & Masson, D. (1981). Photovoltaci effects in Cu$_2$O-Cu cells growing by anodic oxidation. *Solid-St. Electron.*, Vol. 25, 4, 281-283

Garuthara, R. & Siripala, W. (2006). Photoluminescence characterization of polycrystalline n-type Cu$_2$O films. *J. Luminescence*, Vol. 121, 173-178

Ghijsen, J., Tjeng, L.H., Elp, J. V., H. Eskes, Westerink, J., & Sawatzky, G.A. (1988). Electronic structure of Cu$_2$O and CuO. *Phys. Rev.*, Vol. 38, 11322-11330

Guy, A. C. (1972). *Introduction to Material Science (International Student Edition)*, McGraw-Hill, Tokyo

Hames,Y. & San, S. E. (2004).CdO/Cu$_2$O solar cells by chemica deposition. *Sol. Energy*, Vol. 77, 291-294

Harukawa, N., Murakami, S., Tamon, S., Ijuin, S., Ohmori, A., Abe, K. & Shigenari, T. (2000). Temperature dependence of luminescence lifetime in Cu$_2$O. *J. Luminescence*, Vol. 87-89, 1231-1233

Herion, J., Niekisch E. A. & Scharl, G. (1980). Investigation of metal oxide/cuprous oxide heterojunction solar cell. *Sol. Energy Mater.*, Vol. 4, 101-112

Ivill, M., Overberg, M. E., Abernathy, C. R., Norton, D. P., Hebard, A. F., Theoropoulou, N. & Budai, J. D. (2003). Properties of Mn-doped Cu$_2$O semiconducting thin films grown by pulsed-laser deposition. *Solid-St. Electronics*, Vol. 47, 2215-2220

Kaufman, R. G. & Hawkins, R. T. (1984). Defect luminescence of thin films of Cu$_2$O on copper. *J. Electrochem. Soc.*, Vol. 131, 385-388

Mahalingam, T., Chitra, J. S. P., Chu, J. P. & Sebastian, P. J. (2004). Preparation and microstructural studies of electrodeposited Cu$_2$O thin films. *Mater. Lett.*, Vol. 58, 1802-1807

Mahalingam, T., Chitra, J. S. P., Rajendran, S. & Sebastian, P. J. (2002). Potentiostatic deposition and characterisation of Cu$_2$O thin films. *Semicond. Sci. Technol.*, Vol. 17, 565- 570

Mahalingam, T., Chitra, J. S. P., Rajendran, S., Jayachandran, M. & Chockalingam, M. J. (2000). Galvanostatic deposition and characterization of cuprous oxide thin films. *J. Crys. Growth*, Vol. 216, 304-310

Maruyama, T. (1998). Copper oxide thin films prepared by chemical vapor deposition from copper dipivaloylmethanate. *Sol. Energy Mater. Sol. Cells*, Vol. 56, 85-92

Musa, A. O., Akomolafe, T. & Carter, M. J. (1998). Production of cuprous oxide, a solar cell material, by thermal oxidation and a study of its physical and electrical properties. *Sol. Energy Mater. Sol. Cells*, Vol. 51, 305-316

Ogwa, A. A., Bouquerel, E., Ademosu, O., Moh, S., Crossan, E. & Placido, F. (2005). An investigation of the surface energy and optical transmittance of copper oxide thin films prepared by reactive magnetron sputtering. *Acta Materialia*, Vol. 53, 5151-5159

Olsen, L. C., Addis, F. W. & Miller, W. (1981-1983). Experimental and theoretical studies of Cu_2O solar cells. *Sol. Cells*, Vol. 7, 247-279

Papadimitriou, L., Economou N. A. & Trivich, D. (1981). Heterojunction solar cells on cuprous oxide. *Sol. Cells*, Vol. 3, 73-80

Paul, G. K., Nawa, Y., Sato, H., Sakurai, T. & Akimoto, K. (2006). Defects in Cu_2O studied by deep level transient spectroscopy. *Appl. Phys. Lette.*, Vol. 88, 141900

Pollack, G. P. & Trivich, D. (1975). Photoelectric properties of cuprous oxide. *J. Appl. Phys.*, Vol. 46, 163-173

Rai, B. P. (1988). Cu_2O solar cells: a review. *Sol. Cells*, Vol. 25, 265-272

Rakhshani, A.E. (1986). Preparation, charaterestics and photovoltaic proporties of cuprus oxide – a review. *Soild State Electronics*, Vol. 29, No. 1, 7-17

Rakhshani, A. E. & Varghese, J. (1988). Potentiostatatic electrodeposition of cuprous oxide. *Thin Solid Films*, Vol. 157, 87-95

Rakhshani, A. E. & Varghese, J. (1987). Galvanostatic deposition of thin films of cuprous oxide. *Sol. Energy Mater.*, Vol. 15, 237-248

Roos, A., Chibuye, T. & Karlsson, B. (1983). Proporties of oxide copper sufaces for solar cell applications II. *Sol. Energy Mater.*, Vol. 7, 453 467-480

Santra, K., Chitra, C. K., Mukherjee, M. K. & Ghosh, B. (1992). Copper oxide thin films grown by plasma evaporation method. *Thin Solid Films*, Vol. 213, 226-229

Sears, W. M., Fortin, E. & Webb, J. B. (1983). Indium tin oxide/Cu_2O photovoltaic cells. *Thin Solid Film*, Vol. 103, 303–309

Sears,W. M. & Fortin, E. (1984). Preparation and properties of Cu_2O/Cu photovoltaic cells. *Sol. Energy Mater.*, Vol. 10, 93-103

Siripala, W., Perera, L. D. R. D., De Silva, K. T. L., Jayanetti, J. K. D. S., & Dharmadasa, I. M. (1996). Study of annealing effects of cuprous oxide grown by electrodeposition technique. *Sol. Energy Mater. Sol. Cells*, Vol. 44, 251-260

Siripala, W. & Jayakody, J. R. P. (1986). Observation of n-type photoconductivity in electrodeposited copper oxide film electrodes in a photoelectrochemical cell. *Sol. Energy Mater.*, Vol. 14, 23-27

Stareck, J. E. (1937). US Patent #2,081,121

Tanaka, H., Shimakawa, T., Miyata, T., Sato H. & Minami, T. (2004). Electrical and optical properties of TCO–Cu_2O heterojunction devices. *Thin Solid Films*, Vol. 469, 80-85

Tang, Y., Chen, Z., Jia, Z., Zhang, L. & Li, J. (2005). Electrodeposition and characterization of nanocrystalline cuprous oxide thin films on TiO_2 films. *Mater. Lett.*, Vol. 59, 434-438

Tiwari, A.N., Pandya, D.K. & Chopra, K.L. (1987). Fabrication and analysis of all-sprayed $CuInS_2$/ZnO solar cells. *Solar Cells*, Vol. 22, 263-173

Wijesundera, R.P., Hidaka, M., Koga, K., Sakai, M., & Siripala,W. (2006). Growth and characterisation of potentiostatically electrodeposited Cu_2O and Cu thin films. *Thin Solid Films*, Vol. 500, 241-246

Wijesundera, R. P., Perera, L. D. R. D., Jayasuriya, K. D., Siripala, W., De Silva, K. T. L., Samantilleka A. P. & Darmadasa, I. M. (2000). Sulphidation of electrodeposited cuprous oxide thin films for photovoltaic applications. *Sol. Energy Mater. Sol. Cells,*Vol. 61, 277-286

Wijesundera, R. P. (2010). Fabrication of the CuO/Cu_2O heterojunction using an electrodeposition technique for solar cell applications. *Semicond. Sci. Technol.*, Vol. 25, 1-5

Wijesundera, R.P., Hidaka, M., Koga, K., Sakai, M., Siripala, W., Choi, J.Y. & Sung, N. E. (2007). Effects of annealing on the properties and structure of electrodeposited semiconducting Cu-O thin films, *Physica Status of Solidi (b)*, Vol. **244,** 4629-4642

5

Application of Electron Beam Treatment in Polycrystalline Silicon Films Manufacture for Solar Cell

L. Fu

College of Materials Science, Northwestern Polytechnical University, Xian,
State Key Laboratory of Solidification Processing
P. R. China

1. Introduction

Solar cell attracts more and more attentions recently since it transfers and storages energy directly from the sun light without consuming natural resources on the earth and polluting environment. In 2002, the solar industry delivered more than 500 MW per year of photovoltaic generators. More than 85% of the current production involved crystalline silicon technologies. These technologies still have a high cost reduction potential, but this will be limited by the silicon feedstock (Diehl et al., 2005; Lee et al., 2004). On the other hand the so-called second generation thin film solar cells based on a-Si, µc-Si, Cu(In,Ga)(Se,S)2, rare earth or CdTe have been explored(Shah et al.,2005; Li et al.,2004). Crystalline silicon on glass (CSG) solar cell technology was recently developed by depositing silicon film on a glass substrate with an interlayer. It can addresses the difficulty that silicon wafer-based technology has in reaching the very low costs required for large-scale photovoltaic applications as well as the perceived fundamental difficulties with other thin-film technologies (M. A. Green et al., 2004). This technology combines the advantages of standard silicon wafer-based technology, namely ruggedness, durability, good electronic properties and environmental soundness with the advantages of thin-films, specifically low material use, large monolithic construction and a desirable glass substrate configuration.

This Chapter will descript research about the polycrystalline silicon thin film absorber based on CSG technology with high efficiency. Line shaped electron beam recrystallized polycrystalline silicon films of a 20µm thickness deposited on the low cost borosilicate glass-substrate, which are the base for a solar cell absorber with high efficiency and throughput. It is known that the morphology of polycrystalline silicon film and grain boundaries have strong impact on the photoelectric transformation efficiency in the later cell system. Thus, this study concentrates on the influence of recrystallization on the silicon-contact interface and the surface morphology.

2. Experiment methods

Fig. 1 shows the schematic illustration of the silicon solar cell used in this work. The substrate of polycrystalline silicon thin film is Borosilicate glass, which is $10\times10\times0.07cm^3$ in size. A pure tungsten layer of 1.2µm was sputtered on the glass substrate at DC of 500W in an argon atmosphere, which has almost the same thermal expansion coefficient of $4.5 \times 10^{-6} K^{-1}$ as that of the silicon film (Linke et al., 2004; Goesmann et al., 1995). This tungsten interlayer was used as a thermal and mechanical supporting layer for deposition of the silicon film. Nanocrystalline silicon films were then deposited on the tungsten interlayer by the plasma enhanced chemical vapour deposition process (PECVD) within $SiHCl_3$ and H_2 atmosphere. Details of the process were described in References (Rostalsky et al., 2001; Gromball et al., 2004, 2005). The power density used was $2.5 W/cm^2$. The gap in the PECVD parallel plate reactor was 10mm and the substrate temperature was 550°C. The flow rate $H_2/SiHCl_3$ is 0.25 to reduce the hydrogen and chlorine content in the film. Boron trichloride (BCl_3) was added in the gas for an in-situ p-doping. The process pressure was chosen to 350 Pa for the minimized stress. At the above conditions, the deposition rate up to 200nm/min was obtained. After a silicon film of 15-20µm thickness was deposited, a SiO_2 layer of 400nm thickness was deposited on the top of the silicon from $SiHCl_3$ and N_2O within 5 min to prevent balling up.

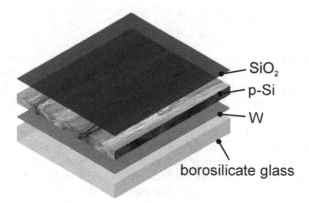

Fig. 1. Structure of thin film silicon solar cell

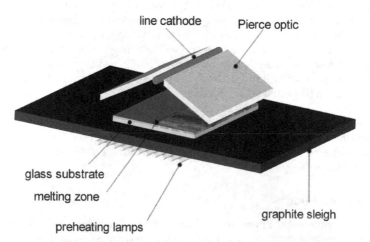

Fig. 2. Schematic of the linear electron beam recrystallization system (Gromball et al., 2005)

The P-doped polycrystalline silicon absorber of 10cm² was melted and recrystallized by a controlled line shaped electron beam (size in 1×100mm²) as described in Fig.2. The appearance of the sample after recrystallization was shown in Fig.3. The samples are preheated from the backside to 500°C within 2 min by halogen lamps. The electron beam energy density applies to the films is a function of the emission current density, the accelerating voltage and the scan speed. The scan speed is chosen to 8mm/s and the applied energy density changes between 0.34J/mm² and 0.4J/mm². To obtain the required grain size, the silicon should be melted and re-crystallized. Therefore, temperature in the electron beam radiation region should be was over the melting point of silicon of 1414°C. The surface morphology of the film, as well as distribution of WSi₂ phase under different energy densities has been investigated by means of a LEO-32 Scanning Electron Microscopy.

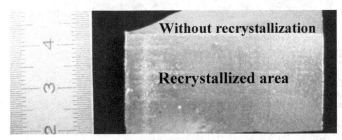

Fig. 3. Appearance of polycrystalline silicon absorber after recrystallization

3. Results and discussion

3.1 Microstructure of the capping layer

The applied recrystallization energy density strongly influences the surface morphology and microstructure of the recrystallized silicon film. With the energy increasing, the capping layer becomes smooth and continuous and less and small pinholes form in the silicon film. Excess of recrystallization energy density leads to larger voids in the capping layer, more WSi₂/Si eutectic crystallites, a thinner tungsten layer and a thicker tungstendisilicide layer.

Fig.4 gives the top view of the polycrystalline silicon film after the recrystallization. The EB surface treatment leads to recrystallization to obtain poly-Si films with grain sizes in the order of several 10µm in width and 100µm in the scanning direction as shown in Fig.5. The polycrystalline silicon films in Fig.4 are EB remelting with four different EB energy densities. Area A was treated with an energy density of 0.34J/mm² (the lowest of the four areas) while area D was treated with an energy density of 0.4J/mm² (highest of the four areas) on the same nanocrystalline silicon layer.

Fig. 4. Top view of the recrystallized silicon film, with increase of applied energy density from the left to the right

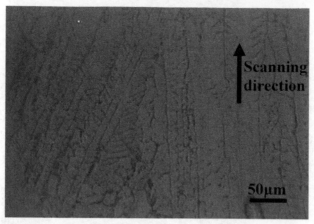

Fig. 5. Grain microstructure of Ploy-Silicon absorber after recrystallization

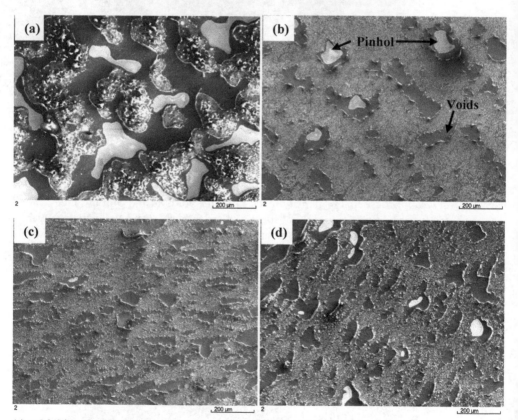

(a) ϵ=0.34J/mm2 ; (b) ϵ=0.36J/mm2 ; (c) ϵ=0.38J/mm2; (d) ϵ=0.4J/mm2

Fig. 6. Surface morphology of the recrystallized silicon layer under different energy density ϵ (Fu et al., 2007)

Fig.6 and Fig.7 show the morphology and microstructure of the EB treated layers. The nanocrystalline silicon is zone melted and recrystallized (ZMR) completely under all the energy chosen in this experiment. It can be seen that after the EB surface treatment, micro-sized silicon grains were formed in all the samples treated under different electron beam energy density ϵ.

The outmost surface was silicon dioxides with some voids and pinholes (bright spots), as shown in Fig.6. Large areas with a rough surface were where the silicon dioxide capping layer (SiO$_2$) existed. The voids (the dark area in Fig.6) in the silicon dioxide capping layer penetrated into the silicon layer with smooth edges. The bright areas were the bottom of the pinholes in which the WSi$_2$ remained.

Influences of the EB energy density on the morphology of deposited films are summarized in Table 1. The energy density influences the surface morphology of the film system strongly. The capping layer exhibited more voids when a lower EB energy density was used, as shown in Fig.6a. The SiO$_2$ capping layer is rougher and appeared as discontinuous droplet morphology in this condition. In addition, large tungstendisilicide pinholes formed due to the lower fluidity and less reaction between the silicon melt and the tungsten interlayer. When the EB energy density was increased, the capping layer becomes smoother and the size of voids was reduced. The number and size of pinholes also became smaller. However, when excess EB energy was applied, the solidification process became unstable and the amount of pinholes increased again. The silicon dioxide capping layer became discontinuous in this case, as shown in Fig. 6d.

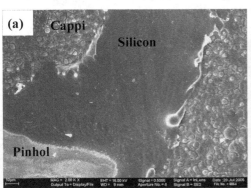 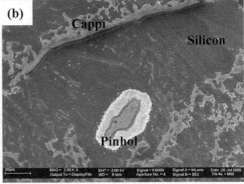

(a) ϵ=0.34J/mm^2 ; (b) ϵ=0.4J/mm^2

Fig. 7. Microstructure of the capping layer and silicon grain under different energy density ϵ (Fu et al., 2007)

It was suggested that the voids are caused by the volume change of the capping layer and the silicon melt during the recrystallization process. Early work [6] suggested that the silicon dioxide in the capping layer could be considered as a fluid with a relatively high viscosity at the EB treatment temperature. For the same amount of silicon, the volume of the solid V_S is about 1.1 times of that of the liquid V_L. Therefore, during solidification process of the silicon melt, the volume increases will produce a curved melt surface. This will generates a tensile stress in the capping layer because of he interface enlargement between the viscous capping layer and the molten silicon. Once the critical strain of the capping layer is surpassed, voids will form in the capping layer. Due to the surface tension of the capping layer and its adhesion to the silicon melt the capping layer also arches upwards and widens the voids. This effect is enhanced by thermal stress and outgassing during the solidification process [5]. As the size, area and viscosity of the SiO_2 layer is affected by the EB energy density, the size and the number of the voids in the capping layer are dependant on the EB energy density as well.

Energy level	SiO₂ capping/ voids	pinholes	$W_{remaining}$/ WSi_2 ratio	WSi₂/Si eutectic
Low (0.34J/mm²)	rough, droplet morphology	High density, biggest (>200µm)	21.7%	fine
Middle (0.36-0.38J/mm²)	smooth, continuous	sporadic, small size (<50µm)	13.3%	coarse
High (0.4J/mm²)	smooth, discontinuous	Low density, bigger(<100µm)	10.5%	coarser and widely spread

Table 1. Influence of the recrystallization energy on the surface morphology of the silicon film system

3.2 Formation of eutectic (WSi₂/Si)

This Chapter gives the details about the formation of Tungstendisilicide (WSi₂). The film system consists of a 20µm thick silicon layer on a 1.2µm thick tungsten film. Tungstendisilicide (WSi₂) is formed at the interface tungsten/silicon but also at the grain boundaries of the silicon. Because of the fast melting and cooling of the silicon film, the solidification process of the silicon film is a nonequilibrium solidification process.

It was claimed that tungstendisilicides were formed in their tetragonal (Hansen, 1958; Döscher et al., 1994) by the solid/solid state reaction and the solid/liquid state reaction between tungsten and silicon according to equation (1) and (2).

$$2Si_{(s)} + W_{(s)} \xrightarrow{\ 700°C-1390°C\ } WSi_{2(s)} \tag{1}$$

$$2Si_{(l)} + W_{(s)} \xrightarrow{\ >1390°C\ } WSi_{2(s)} \tag{2}$$

Formation of the eutectics can be explained using the phase diagram of the Si-W alloy system, as shown in Fig.8. The reactions should start at temperatures above 700°C. The eutectic crystallites (WSi₂/Si) are precipitated from the silicon melt at a eutectic concentration of 0.8 at% W at the eutectic temperature 1390°C in thermal equilibrium. With the temperature increased to above the eutectic temperature (1390°C) for tungsten enriched silicon melt, the WSi₂ layer mainly formed through a solid-liquid reaction and the thickness of the silicide layer increased rapidly. Because 100ms (the FWHM of the electron beam related to the scan speed) were sufficient to generate the tungstendisilicide layer. However, in this experiment, the solidification process of the nanocrystalline silicon was completed within 12.5 seconds for a sample of 10cm² area. Therefore, the solidification process was completed in a nonequilibrium state and the liquid-solid transformation line will divert from equilibrium line shown in Fig.8. At the beginning of the silicon solidification, the formation of tungstendisilicide crystallites will be suppressed by the rapid freezing and followed by the formation of solid silicon. These crystallites start to form just below the liquid-solid transformation temperature, and their growth will be not immediately accompanied by the tungstendisilicide crystallite formation. Therefore, the silicon phase forms dendrites, which grow over a range of temperature like ordinary primary crystallites. Below the eutectic reaction temperature, the remaining melt solidifies eutectically as soon as the melt is undercooled to a critical temperature to allow silicon crystallite growth.

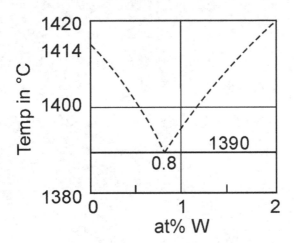

Fig. 8. Phase diagram of the Si-W alloy system in equilibrium (Hansen, 1958)

3.3 Microstructure and distribution of the eutectic crystallites (WSi2/Si) under different recrystallization energy

Tungstendisilicide (WSi_2) was formed at the tungsten/silicon interface but also at the grain boundaries of the silicon throughout all the EB energy density range. A top view scanning electron spectroscopy (SEM) and EDX analysis of the surface region showed that eutectic structure (tungstendisilicide precipitates / silicon) were mainly localized at the recrystallized silicon grain boundaries, as is shown in Fig.9. A typical hypoeutectic structure was found in the exposed silicon layer, which consisted of cored primary silicon dendrites (dendritic characteristic was not very evident) surrounded by the eutectic of the silicon and the tungstendisilicide precipitates. In this eutectic, tungstendisilicide (white areas in the lamellar shape) grew until the surrounding silicon melt had fully crystallized. The eutectic statistically distributed at the primary silicon grain boundaries. The formation and distribution of the eutectic depended on the crystallization and the growth dynamic of the tungsten enriched silicon melt. This is a nonequilibrium solidification process.

The size and the amount of the tungstendisilicide/silicon eutectic depended on the course of the process: when the higher the energy was used in the recrystallization process of the silicon layer, more and large tungstendisilicide crystals grew in the silicon melt. In addition, the WSi2/Si eutectic became coarser at the primary silicon grain boundaries and spread more widely. This was due to the prolonged solidification period for the tungsten enriched silicon melt in the remaining liquid, primarily at the grain boundary. At these sites, the tungstendisilicide crystallites precipitated in the final solidification areas at lower temperature than in case of equilibrium, due to the high tungsten concentration in the

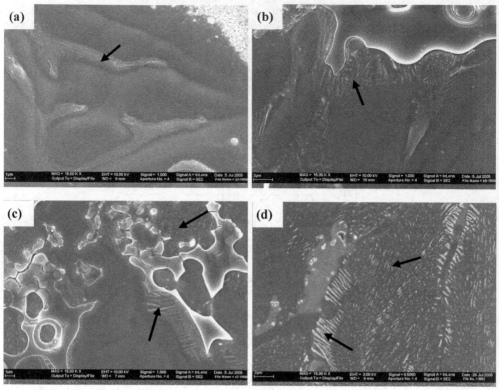

(a) ε=0.34J/mm²; (b) ε=0.36J/mm²; (c) ε=0.38J/mm²; (d) ε=0.4J/mm²

Fig. 9. SEM results of the eutectic structure under different recrystallization energy density ε (Fu et al., 2007)

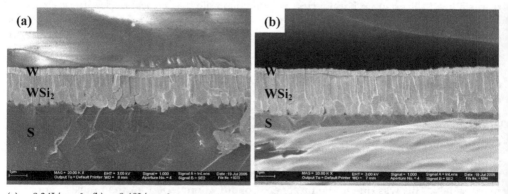

(a) ε=0.34J/mm² ; (b) ε=0.40J/mm²

Fig. 10. Cross section of typical silicon film system under different energy density ε (Fu et al., 2007)

volume. For high EB energy density there was more time for the precipitation and growth of tungstendisilicide and thus more tungstendisilicide crystallites were precipitated at the silicon grain boundaries. The strong tendency of formation of tungstendisilicide at the primary grain boundaries would reduce the efficiency of the solar absorber. Thus a high energy density is not favorable for the recrystallization process.

Fig.10 shows the cross section of a typical resolidified silicon film remelted with different EB energy densities. Tungstendisilicides (WSi$_2$) were formed in the region between the tungsten layer and the silicon layer without relationship to the EB energy density range applied in this research. A thick tungstendisilicide of 2.0-2.86μm exhibited in this experiment. The higher the applied EB energy density, the thicker the tungstendisilicide layer between the tungsten and the silicon layer, the thinner the remaining tungsten layer will be.

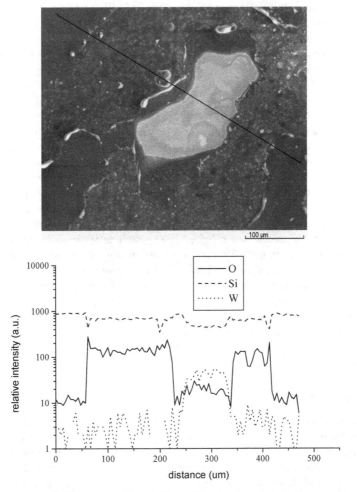

Fig. 11. SEM top view of the recrystallized silicon film in the pinhole area and its EDX line profile mapping results

3.4 Impurities in the recrystallized silicon film

The relatively high chlorine and hydrogen concentrations in the order of 0.5at% lead to outgassing during the recrystallization in completely melting regimes. This effect makes the capping layer arch upwards and widens the voids. Isolated pinholes in the silicon film can be observed. A weak hydrogen chloride peak is detected by mass spectrometry in the base gas atmosphere of the recrystallization chamber. Fig.11 shows an area surrounding a pinhole taken with SEM and the relative element concentrations measured by energy dispersive x-ray analysis (EDX) along the black line. There are no chlorine and hydrogen in the area surrounding a pinhole in the recrystallized film.

4. Summary

This Chapter descried the influence of the applied EB energy density used for the recrystallization process on the surface morphology of the ploy-silicon film system. At a low EB energy density, the voids were formed in the capping layer and the SiO_2 capping layer exhibited a rougher and droplet morphology. With the increase of EB energy density, the capping layer became smooth and the size of the voids decreased. The size and amount of pinholes increased again if the EB energy density was too high. This also led to the formation of larger voids in the capping layer as well as coarser and wider spreading of a WSi_2/Si eutectic crystallite at the grain boundaries.

This Chapter also gave the details about the formation of Tungstendisilicide (WSi_2). The tungstendisilicide precipitates/silicon eutectic structures were mainly localized in at the tungsten/silicon interface but also at the grain boundaries of the silicon throughout all the EB energy density range, as well as the relationship between energy density and microstructure of WSi_2/W areas. Tungstendisilicide forms in its tetragonal by the reaction of tungsten with silicon. WSi_2 improves the wetting and adhesion of the silicon melt but the tungsten layer may degrade the electrical properties of the solar absorber. The formation and distribution of the eutectic depended on the crystallization and the growth dynamic of the tungsten enriched silicon melt. This is a nonequilibrium solidification process.

A tungstendisilicide layer was formed between the tungsten layer and the silicon layer for all EB energy densities used. The higher the applied EB energy density, the thicker the tungstendisilicide layer grows and the thinner the tungsten layer left. It is important to perform the recrystallization process at a moderate energy density to suppress the formation of both WSi_2/Si eutectic and pinholes. In addition, there are no chlorine and hydrogen in the area surrounding a pinhole after recrystallization because of outgassing during the solidification.

5. Acknowledgements

The author would like to thank Prof. J. Müller and Dr. F. Gromball of Technische Universit.t Hamburg-Harburg in Germany for providing experimental conditions and interesting discussion, and also remember Prof. J. Müller with affection for his human and scientific talents. This research was financially supported by the German Federal Ministry for the Environment, Nature Conservation and Nuclear Safety under contract #0329571B in collaboration with the Hahn Meitner Institute (HMI), Berlin-Adlershof, Department for Solar Energy Research. The author was financially supported by China Scholarship Council

(CSC) and the Research Fund of the State Key Laboratory of Solidification Processing (NWPU), China (Grant No. 78-QP-2011).

6. References

Diehl W., Sittinger V. & Szyszka B. (2005). Thin film solar cell technology in Germany. *Surface and Coatings Technology*, Vol.193, No. 1-3, (April 2005), pp.329-334, ISSN: 0257-8972

Döscher M., Pauli M. and Müller J. (1994). A study on WSi_2 thin films, formed by the reaction of tungsten with solid or liquid silicon by rapid thermal annealing. *Thin Solid Films*, Vol.239, No. 2, (March 1994), pp.251-258, ISSN: 0040-6090

Dutartre D. (1989). Mechanics of the silica cap during zone melting of Si films. *Journal of Apply Physics*, Vol.66, No. 3, (August 1989), pp.1388-1391, ISSN: 0021-8979

Fu L., Gromball F., Groth C., Ong K., Linke N. & Müller J. (2007). Influence of the energy density on the structure and morphology of polycrystalline silicon films treated with electron beam. *Materials Science and Engineering B*, Vol.136, No. 1, (January 2007), pp.87–91, ISSN: 0921-5107

Green M. A., Basore P. A., Chang N., Clugston D., Egan R., Evans R. Hogg D., Jarnason S., Keevers M., Lasswell P., O'Sullivan J., Schubert U., Turner A., Wenham S. R. & Young T. (2004). Crystalline silicon on glass (CSG) thin-film solar cell modules. *Solar Energy*. Vol.77, No. 6, (December 2004) , pp.857-863, ISSN: 0038-092X

Goesmann F. & Schmid-Fetzer R. (1995). Stability of W as electrical contact on 6H-SiC: phase relations and interface reactions in the ternary system W-Si-C. *Materials Science and Engineering B*, Vol. 34, No. 2-3, (November 1995), pp.224-231, ISSN: 0921-5107

Gromball F., Heemeier J., Linke N., Burchert M. & Müller J. (2004). High rate deposition and in situ doping of silicon films for solar cells on glass. *Solar Energy Materials & Solar Cells*, Vol.84, No. 1-4, (October 2004), pp.71-82, ISSN: 0927-0248

Gromball F., Ong K., Groth C., Fu L., Müller J., Strub E., Bohne W. & Röhrich J. (2005). Impurities in electron beam recryatallised silicon absorbers on glass, *Proceedings of 20th European Photovoltaic Solar Energy Conference and Exhibition*, Barcelona, Span, July, 2005.

Hansen M. (1958). Constitution of binary alloys, In: *Metallurgy and Metallurgical Engineering Series*, Kurt Anderko, pp.100-1324, McGraw-Hill Book Company, ISBN-13: 978-0931690181, ISBN-10: 0931690188, London

Lee G. H., Rhee C. K. & Lim K. S. (2006). A study on the fabrication of polycrystalline Si wafer by direct casting for solar cell substrate. *Solar Energy*, Vol.80, No. 2, (February 2006), pp.220-225, ISSN: 0038-092X

Li B. J., Zhang C. H. & Yang T. (2005). *Journal of Rare Earths*. Vol.23, No. 2, (April 2005), pp.228-230, ISSN: 1002-0721

Linke N., Gromball F., Heemeier J. & Mueller J. (2004). Tungsten silicide as supporting layer for electron beam recryatallised silicon solar cells on glass, *Proceedings of 19th European Photovoltaic Solar Energy Conference and Exhibition*, Paris, France, July, 2004.

Rostalsky M. & Mueller J. (2001). High rate deposition and electron beam recrystallization of
    silicon films for solar cells. *Thin Solid Films*, Vol.401, No. 1-2, (December 2001),
    pp.84-87, ISSN: 0040-6090

Shah A. V., Schade H., Vanecek M., Meier J., Vallat-Sauvain E., Wyrsch N., Kroll U., Droz
    C. & Bailat J. (2004). Thin-film silicon solar cell technology. *Progress in
    Photovoltaics: Research and Applications*. Vol.12, No. 2-3, (March 2004), pp.113-142,
    ISSN: 1099-159X

6

Crystalline Silicon Thin Film Solar Cells

Fritz Falk and Gudrun Andrä
Institute of Photonic Technology
Germany

1. Introduction

In the last few years the marked share of thin film solar cells increased appreciably to 16.8% (in 2009). The main part of that increase refers to CdTe modules (9.1%) followed by silicon thin film cells, that is amorphous silicon (a-Si) cells or tandem cells consisting of a-Si and nanocrystalline silicon (μc-Si). For a review on thin film solar cells in general see (Green, 2007) and on a-Si/μc-Si cells see (Beaucarne, 2007). The a-Si cells suffer from a low efficiency. In the lab the highest efficiency up to now is 10.1% on 1 cm² (Green et al., 2011), whereas in the industrial production modules reach about 7%. In order to achieve the required electronic quality of hydrogenated amorphous silicon (a-Si:H), low deposition rate (max. 50 nm/min) PECVD (plasma enhanced chemical vapour deposition) is used for deposition which makes production more expensive as compared to CdTe modules. This is even worse for the layer system in a-Si/μc-Si tandem cells for which the more than 1 μm thick nanocrystalline μc-Si layer is deposited by PECVD, too, however with much lower deposition rates in the 10 nm/min range. Cells consisting just of μc-Si reached 10.1% efficiency (Green et al., 2011), just as a-Si-cells, whereas tandem cells arrived at 11.9%, both for lab cells, whereas in production the results are below 10%. The low deposition rate combined with the limited efficiency, make these cells not too competitive compared to CdTe cells, which, at lower cost, reach 11% in industrial production, or to CIGS (Copper-indium-gallium-diselenide) cells with similar efficiencies.

As an alternative, polycrystalline (grains in the μm range) or multicrystalline (grains >10 μm) silicon thin film solar cells receive growing interest (Beaucarne et al., 2006). The present paper reviews the status of these cells, and on the other hand gives details of laser based preparation methods, on which the authors have been working for many years.

Both types, poly- and multicrystalline silicon thin film cells, are prepared by depositing amorphous silicon followed by some crystallization process. One main advantage of the crystallization process is that the electronic quality of the virgin a-Si is not important. Therefore high rate deposition processes such as electron beam evaporation or sputtering can be used which are much less expensive as compared to low rate PECVD. In case of sputtering doped thin films can be deposited by using doped sputtering targets, whereas in electron beam evaporation the dopands are coevaporated from additional sources. So, in these deposition processes the use of toxic or hazardous gases such as silane, phosphine or diborane is avoided, reducing the abatement cost.

Polycrystalline silicon layers for solar cells can be prepared in a single crystallization step. The layer system containing the doping profile is deposited in the amorphous state and is crystallized in a furnace to result in grains about 1 μm in size. This process had been industrialized by the company CSG and is described in Sect. 2. In the lab CSG reached 10.4% efficiency on 90 cm² minimodules (Keevers et al., 2007). Alternatively pulsed excimer laser melting and solidification can been used, which is a standard process in flat panel display production (Sect. 2).

Preparation of multicrystalline silicon thin film solar cells with grains exceeding 10 μm in size is under investigation. This topic is extensively dealt with in Sect. 3. Usually a two-step preparation scheme is used. In a first step a multicrystalline thin seed layer with the desired crystal structure is prepared (Sect. 3.3), which in a second step is epitaxially thickened (Sect. 3.4). For both, seed layer preparation and epitaxial thickening, different processes have been tested. There are, however, attempts to crystallize the complete layer stack of a thin film solar cell in one electron beam melting step (sect. 3.2). The idea for the multicrystalline cells is that in the large grains recombination is reduced, if the crystal quality is high enough, so that the efficiency should exceed that of cells with μm sized grains. Particularly, if the ratio of grain size to layer thickness is large (e.g. 50), such as in multicrystalline wafer cells, a similar efficiency potential is expected. This would require 100 μm large grains for 2 μm thick silicon layers. The preparation methods for large grained multicrystalline silicon layers divide in low and high temperature processes. The high temperature processes are rather straight forward for producing large grains (Beaucarne et al., 2004). However, temperature resistant substrates are required which are expensive. Much more demanding are preparation methods working at temperatures endured by low cost substrates such as glass. One such method is diode laser crystallization. The epitaxial thickening processes, as well, divide in high and low temperature processes with the same drawbacks and advantages. Several methods are presented in Sect. 3. The result of seed layer preparation as well as epitaxial thickening, via melt or in the solid state, depends on temperature history and is explained by the kinetics of phase transformation. The basic notions of this theory as far as they are important for silicon thin film solar cell preparation, are summarized in Sect. 5. Post-crystallization treatments such as rapid thermal annealing and hydrogen passivation are explained in Sect. 4.

Even single crystalline silicon thin film cells have been prepared by a transfer process starting from a silicon wafer from which a layer is detached and epitaxially thickened to several 10 μm thickness (Reuter et al., 2009; Brendel, 2001; Brendel et al., 2003; Werner et al., 2009) to reach an efficiency of 17%. This type of cells, which are much thicker than the poly- and multicrystalline silicon thin film cells, and which cannot be prepared in the typical sizes of thin film technology such as > 1 m², is not the topic of this paper.

2. Polycrystalline silicon thin film solar cells: 1 μm grains

Polycrystalline silicon thin film solar cells in superstrate configuration have been fabricated industrially for some years by the company CSG in Thalheim, Germany. These are the only cells with grains above 1 μm ever fabricated industrially. The preparation steps are as follows (Green et al., 2004). On a borosilicate glass substrate spherical glass beads are deposited, which finally are responsible for light trapping. Then an about 70 nm thick SiN antireflection and barrier layer is deposited by PECVD. On top about 1.5 μm amorphous silicon (a-Si:H) is deposited again be PECVD including the final doping profile n^+pp^+. The silicon layer is crystallized in the solid state in an 18 h furnace annealing step at about 600°C during which grains of about 1 μm in size form. To activate the dopants a rapid (2 min) thermal annealing step at 900°C follows. The silicon layers are passivated by a hydrogen plasma treatment. Finally rather demanding structuring and contacting processes follow. In production, modules 1x1.4 m^2 in size reached about 7% efficiency. In the lab 10.4% efficiency were achieved on 92 cm^2 minimodules (Keevers et al., 2007). The production was stopped, probably because of the high cost PECVD deposition, which was used because the method was the only one available for silicon deposition in the m^2 range. In the lab, high rate electron beam evaporation was tested as an alternative which delivered minimodules with the efficiency of 6.7%, similar to that of the industrially produced modules (Egan et al., 2009; Sontheimer et al., 2009).

The grain size originating in the furnace anneal is dictated by the interplay of crystal nucleation within the amorphous matrix and growth of the nuclei (see Sect. 5). One can influence both processes by the temperature of the annealing step. Practically, however, there is not much choice. At lower temperature the annealing time required for complete crystallization would reach unrealistic high values so that this is not possible in production. Higher temperatures are not endured by the glass substrate for the time span needed for crystallization. Even at 600°C 18 h are required for crystallization and high temperature resistant borosilicate glass has to be used instead of a much cheaper soda lime glass.

As an alternative for the furnace crystallization pulsed excimer laser crystallization via the melt is a process industrially used in flat panel display production. For this application, however, rather thin films (<100 nm) are required and the resulting grain size is rather small, typically below 1 µm. In the context of solar cell preparation requiring films thicker than 1 µm this method has been mentioned only rarely (Kuo, 2009).

3. Multicrystalline silicon thin film solar cells: grains > 10 µm

3.1 Basic considerations

As mentioned in the last paragraph, grains larger than about 1 µm cannot be prepared by direct deposition of crystalline silicon, nor by solid phase crystallization of a-Si nor via melting a-Si by short laser pulses. Large grains can be produced from the melt only if the melt is cooled below the equilibrium melting point slowly so that the melt stays long enough in a region of low nucleation rate and there is time enough for the few nucleating crystallites to grow to large size. Low cooling rate means low heat flow into the substrate following from a low temperature gradient in the substrate. This can be achieved if the melting time of the silicon layer is larger than in excimer laser crystallization, i.e. much larger than 100 ns. To reach longer melting times the energy for melting has to be delivered on a longer time scale. For energy delivery scanned electron beams or scanned laser beams have been used. However, the longer melting time has the consequence, that dopand profiles, introduced into the virgin a-Si for emitter, absorber, and back surface field, get intermixed due to diffusion. Typical diffusion constants in liquid silicon are in the 10^{-4} to 10^{-3} cm^2/s range (Kodera, 1963) so that dopands will intermix over a distance of 1 µm within 10 to 100 µs. Nevertheless a one-step crystallization procedure for a solar cell layer system has been done by electron beam melting, discussed in Sect. 3.2. Alternatively a two-step procedure has been used. In a first step a thin seed layer is crystallized to large grains from a-Si by laser irradiation. In a second step the seed is thickened epitaxially. Seed and epitaxial layer can be differently doped so that the seed can act as the emitter and the epitaxial layer as the absorber of the solar cell. Alternatively, the seed may act as a highly doped back surface field layer with the epitaxial layer acting as a moderately doped absorber. The emitter is generated on top in a third preparation step.

An important issue in any of the mentioned preparation steps is the choice of the substrate. This choice depends on the thermal load the substrate experiences during the silicon crystallization process. Plastic substrates are not useful for any of the processes described in Sect. 3 since the substrate temperature well exceeds 200°C. One usually divides the crystallization methods into low temperature processes for which glass can be used as a substrate and high temperature processes for which glass is not sufficient. Instead, ceramics (e.g. alumina) or graphite has been used. These substrate materials, however, are much more expensive than glass so that the economic consequences for the high temperature routes are not so pleasant.

Typically, in high as in low temperature processes some barrier layer is used to prevent the diffusion of foreign atoms from the substrate material into the silicon layer during the processing steps. The barrier layer has to fulfil different requirements except of its main purpose. First of all it has to withstand liquid silicon, i.e. it should not decompose or react with the silicon melt. Moreover, it should not release gases which would blow off the silicon layer. Then it should be well wetted by liquid silicon. Otherwise the silicon film during melting could dewet to form droplets. This latter requirement is the reason that SiO_2 is not useful as a barrier layer. Silicon nitride or silicon carbide are better suited. However, if deposited by PECVD the layers contain too much hydrogen which is released during silicon melting so that the silicon films are destroyed. According to our experience sputtered silicon nitride is well suited if prepared correctly.

3.2 Single step layer preparation - electron beam crystallization

As mentioned in Sect. 3.1 silicon solar cell absorbers in substrate configuration have been prepared by electron beam crystallization in a one step process (Gromball et al., 2004; Amkreuz et al., 2009). On a glass substrate with a barrier layer (e.g. SiC) 7 to 15 µm of p-doped (10^{17} cm^{-3} B) nanocrystalline silicon was deposited by high rate (up to 300 nm/min) PECVD from trichlorosilane. This layer was crystallized by scanning a line shaped electron beam (15 cm x 1 mm). At a scanning rate of 1 cm/s a beam energy density of 500mJ/cm^2 has been used so that any position is treated for about 0.1 s. The resulting grain size is in the mm range. To get a solar cell a 30 nm thick n-doped a-Si heteroemitter was deposited onto the crystalline absorber by PECVD. The maximum solar cell parameters achieved so far were j_{sc} = 12.4 mA/cm^2, V_{oc} = 487 mV, and an efficiency of 3.5% (Amkreuz et al., 2009). Obviously the absorber doping is too high and a back surface field is missing. Work is ongoing to improve these cells.

3.3 Two-step process - seed preparation

In the two-step preparation method first a thin seed layer with the desired crystal structure is prepared which can be used as a back surface field layer or as emitter in the final solar cell. The absorber is then prepared by epitaxial thickening of the seed. In case of a cell in superstrate configuration (illumination through the glass), the seed layer should be rather thin. This is to reduce light absorption in the seed which is highly doped (as emitter or as back surface field layer) and shows only low photovoltaic activity. Two seed preparation methods have been investigated: aluminium induced crystallization (Fuhs et al., 2004) as well as laser crystallization.

3.3.1 Aluminum induced crystallization for seed preparation

Aluminum induced crystallization (AIC) works as follows: On to the substrate an aluminum layer is deposited by sputtering or evaporation. On top follows an amorphous silicon layer.

When the Al/a-Si layer system is heated (350°C...550°C below the eutectic temperature of the Al-Si system at 577°C) a layer exchange process takes place combined with silicon crystallization, which is completed, at 500°C, in about 30 min. (Pihan et al., 2007). Finally, a crystalline silicon layer rests on the glass and is covered by an aluminium layer, which may contain silicon islands. The silicon layer is highly p-doped typically by 10^{19} cm^{-3} Al (Antesberger et al., 2007). It has been reported that the details of the process and the properties of the final silicon layer depend on the thickness of an aluminum oxide layer which was present between Al and a-Si before the tempering step. Typical resulting silicon grain sizes are in the range of 10 μm. The preferred grain orientation is (100) but other orientations occur as well (Schneider et al., 2006a). Typical layer thicknesses are 300 nm for Al and 375 nm for Si (Fuhs et al., 2004), which is a bit high for seed layers. However, even silicon films thinner than 100 nm have been crystallized by AIC (Antesberger, 2007). Some work has been done to understand the thermodynamics and the kinetics of the process (Wang et al., 2008; Sarikov et al., 2006; Schneider et al., 2006b). It seems that silicon diffuses through the thin alumina layer into the aluminum where it preferably further diffuses towards the glass along the aluminum grain boundaries. When aluminum gets supersaturated by silicon, nucleation of silicon crystallites starts preferably at the interface to the glass substrate. The driving force for the process is the free energy difference between metastable amorphous and absolutely stable crystalline silicon. Finally, the a-Si completely has diffused through the aluminum which then rests on top. Before the crystalline silicon layer can be used as a seed, the aluminum layer has to be removed, e.g. by wet chemical etching using HCl. Challenging is the removal of the silicon islands included in the aluminum layer and of the aluminum oxide film. The removal of both is crucial for good epitaxy (Rau et al., 2004). The inverse process with the starting sequence glass/a-Si/Al and the final sequence glass/Al/c-Si works as well (Gall et al., 2006). It has some advantages for cells in substrate configuration, e.g. that a Al back contact is formed automatically. However, the Al/Si contact has the consequence that any further processing steps, e.g. epitaxy, cannot be performed above the eutectic temperature of the Al-Si system of 577°C. For this reason the inverse process was abandoned.

There has been done a lot of work on silicon crystallization by other metals, e.g. Au, Ni, but these methods did not find application in solar cell preparation.

3.3.2 Laser crystallization for seed preparation

To get large silicon crystals by laser crystallization the beam of a cw laser is scanned so that the irradiation time at each position is in the ms range, much larger than during pulse laser irradiation mentioned in Sect. 2. Under these conditions the temperature gradient and therefore the heat flow in the substrate is low so that the melt undercools only slowly, nucleation rates are low, and nucleated crystals have time enough to grow to large sizes (see Sect. 5). First results on this method date back to the late 1970ies (Gat et al., 1978; Colinge et al., 1982). At these times laser crystallization was performed for applications in microelectronics. Therefore amorphous silicon on wafers covered by oxide was used as starting material. The only available well suited lasers were argon ion lasers emitting green light at 514 nm wavelength with a total power of up to 15 W. Typically a circular Gaussian beam with diameter in the 40 μm range was scanned across the sample. At a scanning rate of 12.5 cm/s already in 1978 grains 2x25 μm in size were produced (Gat et al., 1978). Due to the high thermal conductivity of the wafer substrate a rather high power density is needed for melting and crystallization in this case. Only later glass was discovered as a useful substrate for thin film transistor applications (Michaud et al., 2006) as well as for solar cells (Andrä et al., 1998; Andrä et al. 2000). On glass with low thermal conductivity power densities of about 20 kW/cm² are needed at scanning speeds of several cm/s. Due to the limited laser power the spot diameter was limited to about 100 µm.

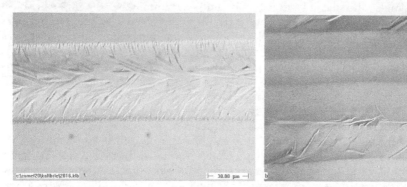

Fig. 1. Optical micrograph of a silicon layer crystallized by scanning the circular beam of an argon ion laser. Left: single scan; right: overlapping scans

Fig. 1 (left) shows an optical micrograph of a single trace produced by scanning a circular Ar ion laser beam. At the rim very fine crystals were produced. There the laser power just was able to generate a temperature a bit above the melting point of a-Si, which is well below the melting point of crystalline silicon (see Sect. 5.1). In the rim region a strongly undercooled melt is generated which immediately (that is must faster than the irradiation time) crystallizes to fine grained (about 100 nm) silicon. Towards the center of the trace the power density increases so that the temperature gets higher, the undercooling gets lower, and a bit larger grains solidify. In the central part the laser power is high enough to produce a silicon melt above the equilibrium melting point of crystalline silicon (1412°C). There solidification occurs only when the laser beam already has passed. The slowly undercooling liquid silicon is in contact with the small crystallites of the rim region which crystallized earlier. From these, lateral epitaxial growth takes place. The crystallization direction coincides with the temperature profile following the scanned laser beam. Those of the many nuclei are successful in epitaxy for which the fastest crystallographic growth direction coincides with the temperature gradient. Therefore, a selection mechanism is active and only few of the potential nuclei grow. As a consequence large grains form several 10 µm wide and over 100 µm long. To get not just one crystalline trace but a completely crystallized area, one just has to scan the laser beam in overlapping rows (Fig. 1, right). In the second row the laser beam remelts part of the previous row with the consequence that now the melt is in contact with the large grains produced in the previous row. Therefore large crystals are already present for lateral epitaxy to occur. In this way large areas covered by large grains can be produced. Defect population in films generated in this way has been investigated (Christiansen et al., 2000). The dislocation density was rather low. Grain boundaries are mostly Σ3 and Σ9 twin boundaries which are expected to be not active electrically. The grain orientation is at random with no preferential texture.

Later on, for crystallization the argon ion laser was replaced by a solid state cw Nd:YAG laser, emitting green light of 532 nm wavelength after frequency doubling. Similar results were obtained with this laser type (Andrä et al., 2005a). Both, argon ion as well as Nd:YAG lasers, have rather limited power so that it is impossible to crystallize seed layers for large area solar cells in an industrial environment. For example, a 1 m² module would require many hours laser treatment. Therefore, when looking for high power lasers we ended up with diode lasers, emitting in the near infrared. However, the absorption coefficient of a-Si for 806 nm radiation, the shortest wavelength available for high power diode lasers, at room temperature is only about 0.3 µm⁻¹, as compared to 25 µm⁻¹ for green light. Fig. 2 shows the absorption of 806 nm radiation in amorphous silicon (electron beam deposited, hydrogen free) as calculated from optical properties (n and k) measured from room temperature up to 600°C and extrapolated up to 1000°C. The maxima and minima are due to interference effects in the silicon layer.

Obviously there exists a problem for thin films, particularly at room temperature. In thin films, only a small amount of the incoming radiation is absorbed at room temperature. Therefore, to heat the silicon film, a rather high power density is needed. When heating started successfully then the absorption increases and a run-off sets in which is only limited after melting, when the reflectivity jumps up. So the process has some inherent instability, which can be handled only when one preheats the substrate to about 600°C so that laser heating starts at a higher absorption already. The substrate heating has another positive effect, namely to reduce the cracking tendency of the glass substrate, for which we use a borosilicate glass (Schott boro 33) with a thermal expansion coefficient very near to that of silicon. Work using diode lasers for crystallization started 2006 (Andrä et al., 2006).

For our seed layer crystallization we use LIMO line focus lasers (806 nm wavelength, 13 mm x 0.1 mm focus and 30 mm x 0.1 mm focus) with maximum power density of up to 25 kW/cm² (Andrä et al., 2006), allowing for scanning speeds up to several cm/s. Fig. 3 shows an EBSD map of a crystallized region demonstrating large grains in the 100 µm range in 450 nm thick films. With the diode laser we can go down to 100 nm thin films. In these the grains size is in the 30 µm range. A further problem with thin films is dewetting. This means that holes form when the silicon film is liquid. It even happens that the holes grow to large sizes and only a part of the substrate is covered by silicon. Dewetting can be reduced if the wetting angle of liquid silicon on the substrate is low. This can be influenced by the barrier layer on the glass substrate.

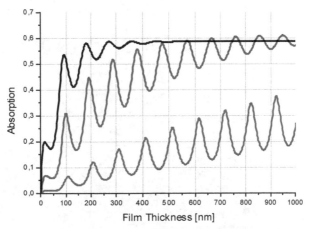

Fig. 2. Absorption of 806 nm diode laser radiation in an amorphous silicon thin film on glass as depending on film thickness. Film temperature 20°C (blue), 600°C (red), and 1000°C (black).

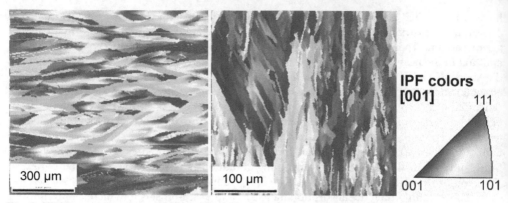

Fig. 3. EBSD map (inverse pole figure) of diode laser crystallized seed layers 450 nm (left) and 110 nm (middle) thick. Color code for grain orientation is shown on the right.

Concerning the throughput, laser companies are just developing line focus diode lasers with long lines (Lichtenstein 2010) which would allow crystallization of a 1 m² module within minutes.

If seed layers thinner than 100 nm are to be crystallized diode lasers cannot be used due to too low absorption even when preheated. We tested a pulsed green laser (JenLas ASAMA) emitting 515 nm wavelength radiation (Andrä et al., 2010). This laser has a line focus up to 100 mm long and 5 to 10 μm wide and it delivers 600 ns pulses at a repetition rate of up to 80 kHz. At a fluence of about 1.2 J/cm² the sample was shifted 1.5 μm between subsequent pulses. In this way 60 nm thin seed layers were crystallized without any preheating with resulting grains several μm wide and several 10 μm long (Fig. 4). Obviously, the melt generated during each laser pulse solidifies by lateral epitaxy so the grains generated by the previous pulse grow stepwise. Finally long grains form, which continue over many pulses. Since the width of the melt is 5 μm in our case and the melt exists for a time interval in the several μs range, the solidification speed is in the m/s range. This value is near the maximum following from solidification kinetics (see Sect. 5).

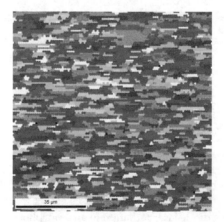

Fig. 4. EBSD map (inverse pole figure) of pulse laser crystallized seed layers 60 nm thick.

3.4 Two step process - epitaxial thickening

In the two step preparation method on top of the multicrystalline seed layer the absorber of the solar cell is prepared by epitaxial growth. Several methods have been used which can be classified into direct epitaxial deposition and deposition as amorphous silicon followed by epitaxial crystallization, either in the solid state by furnace or by laser annealing or via laser melting. Particularly in the cases without melting the cleanliness of the interface between crystalline seed and amorphous silicon to be epitaxially crystallized is an issue. Any contaminants present, even small amounts of a monolayer, will jeopardize epitaxial crystallization or at least increase the amount of extended defects in the epitaxial layer appreciably. First of all, any silicon oxide has to be removed from the seed surface. This can be achieved by HF. A 2% to 5 % solution in water is most useful. Success can be observed by the naked eye. When HF has removed the oxide the silicon surface gets hydrogenated which makes the surface hydrophobic and the etching solution dewets, i.e. forms droplets. Then the HF solution can be blown off by nitrogen. The hydrogenated surface state remains stable in ambient air at room temperature for about 1 h so that there is time enough to introduce the sample into a deposition chamber for a-Si deposition. However, other possible contaminants are not so easily removed. It turned out as useful to start with an RCA cleaning step before HF treatment. The RCA step removes e.g. organic contaminants.

3.4.1 Direct epitaxial deposition

The simplest epitaxial thickening procedure is direct epitaxial deposition of silicon on top of the seed layer. Several processes have been investigated in the past, high temperature CVD and, at intermediate temperature, electron beam evaporation, ECRCVD, and hot wire CVD.

The high temperature route has been reviewed recently (Beaucarne et al., 2004). The highest efficiency reached so far with this method is 8% (Gordon et al., 2007). On an alumina substrate seed layers were prepared by aluminium induced crystallization. Epitaxial thickening for the p-doped absorber with rates up to 1.4 μm/min was done by thermal CVD at 1130°C from trichlorosilane. The final emitter was prepared by phosphorus diffusion, or an a-Si heteroemitter was deposited by PECVD. Corresponding to the seed layer the grain size in the absorber is several 10 μm. It is expected that the efficiency is not so much limited by the grain size but by intragrain defects, which have been thoroughly investigated (van Gestel et al., 2009).

Even higher efficiencies of 11.1% were reached on seed layers crystallized by lamp heater zone melting on graphite and high temperature epitaxy for absorber growth (Kunz et al., 2008). The high temperature process has the advantage that it works on any grain orientation of the seed. However, high temperature resistant substrates such as alumina, silica, glass ceramics, or graphite are needed, which are not very feasible for large scale production.

At intermediate temperature both, electron beam evaporation, partly modified by ion assisted deposition, or ECR-CVD (electron cyclotron resonance CVD) has been tested for epitaxy on AIC seed layers. ECR-CVD was successfully applied at 585°C substrate temperature (Rau et al., 2004). However, epitaxy worked well only on (100)-oriented grains, which is the most common orientation following from AIC, but not the only one. At 670°C epitaxy by hot wire CVD worked on any grain orientation with a rate of 100 nm/min. Ion assisted deposition, that is electron beam evaporation plus some ionization of the silicon atoms, was tested for epitaxy as well. For the deposition a temperature ramp was carefully optimized with maximum temperature below 700°C. The deposition rate was 300 nm/min. The highest achieved open circuit voltage of solar cells was 453 mV (Straub et al., 2005). Direct epitaxy during electron beam evaporation at 550°C substrate temperature has successfully been demonstrated (Dogan et al., 2008). Solar cells prepared with this process reached 346 mV open circuit voltage and 2.3% efficiency, which is a bit low as compared to the values achieved by other methods.

3.4.2 Solid phase epitaxy in furnace

Technically the most simple way to achieve epitaxial growth is to deposit first an amorphous layer on top of the cleaned seed layer, and then to epitaxially crystallize the layer by furnace annealing in the solid state. The layer to be crystallized can already contain the desired doping profile which remains during the annealing step. The main critical point with this simple procedure is that not only an epitaxial crystallization front moves into a-Si, but also spontaneous nucleation will occur within a-Si followed by growth of crystallites. So there exists a competing process to the desired epitaxy. The question arises, which of the two succeeds. The speed of the epitaxial front of course depends on temperature (described by Jackson-Chalmers equation, see Sect. 5.1) and so does nucleation, described by classical nucleation theory (Sect. 5.2), and growth of nuclei, the latter phenomena described together by Avrami-Mehl equation (Sect. 5.4). An important point, which makes SPE possible, is that, if no nuclei pre-exist in the amorphous matrix, nucleation does not start immediately. Instead it needs some time, called time lag of nucleation, until a stationary population of nuclei evolves (Sect. 5.3). Only after that time lag the stationary nucleation rate applies at fixed temperature, described by classical nucleation theory, and crystal nuclei appear. So any successful epitaxy relies on the time lag of nucleation. The thickness of an epitaxially crystallized layer is just given by the time lag of nucleation times the speed of the epitaxial crystallization front. After the time lag, in the virgin amorphous silicon crystalline nuclei of random orientation appear resulting in fine grained material, such as is generated by direct furnace crystallization (see Sect. 2) without seed. For successful epitaxy one has to make sure that within the amorphous phase there are no nuclei present which could form during deposition already.

In the last few years we developed the technique of SPE on diode laser crystallized seed layers on borosilicate glass substrates (Andrä et al., 2008a; Schneider et al., 2010). The virgin a-Si layers including a doping profile were deposited at high rate (typically 300 nm/min) by electron beam evaporation at a substrate temperature in the 300°C range. At that temperature no nuclei form within a-Si. The layer system was then annealed in a furnace under ambient air. To control the progress of crystallization, an in situ measurement technique was installed. For this purpose, the beam of a low power test laser was sent through the sample. The transmitted intensity was monitored by a photocell. Since a-Si has a different optical absorption from c-Si, the progress of crystallization can be monitored easily. In particular, the crystallization process is complete when the transmission does not change any more. Fig. 5 shows a transmission electron micrograph of a cross section of an epitaxially thickened silicon film.

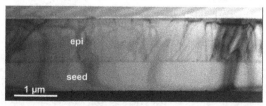

Fig. 5. Transmission electron microscopic cross section image of a film epitaxially thickened by furnace annealing.

In summary we could epitaxially crystallize up to 1.6 μm of a-Si at a temperature of 630°C within 3 h. The epitaxial quality as determined by EBIC was best in (100) oriented grains and worst in (111) grains. Moreover, the epitaxial crystallization speed depends on orientation and on the doping level. Higher doped layers crystallize faster. Solar cells prepared on these layers reached an efficiency of 4.9% after hydrogen passivation (Schneider et al., 2010). By TEM cross section investigations it was shown that the seed layers contain only very few extended defects such as dislocations, whereas the epitaxial layer contains much more. It seems that the cleaning procedure of the seed surface prior to a-Si deposition is crucial for good epitaxial quality. At least the dislocation density in the epitaxial layer could be reduced by an additional RCA cleaning step before removal of oxide by HF. However, this did not reflect in the achieved solar cell efficiencies.

3.4.3 Layered laser crystallization

The epitaxy method of layered laser crystallization has been developed in our group years ago (Andrä et al. 2005b, Andrä et al., 2008a). The principle is simple. During deposition of a-Si on top of the seed layer excimer laser pulses are applied repeatedly, which melt the newly deposited a-Si and a bit of the crystalline silicon beneath so that after each pulse epitaxial solidification occurs. Again, the layer thickness to be crystallized by one laser shot is limited by a competing nucleation process in the undercooling melt after the laser pulse. According to our experience about 200 nm of a-Si can be epitaxially crystallized by one laser pulse. The typical laser fluence needed is 550 mJ/cm². However, when during the whole thickening process the thickness of the crystalline layer beneath the newly deposited a-Si increases from the initial seed layer (say 200 nm) to the final absorber thickness (say 2 μm) the laser parameters or the thickness of the newly deposited a-Si have to be adjusted so that the laser pulse just melts the a-Si and bit of c-Si beneath. This adjustment is necessary because the thermal properties of glass, c-Si, and a-Si differ so that the temperature profiles change during the process if the laser energy would be kept constant. In the layered laser crystallization process epitaxy works independently of the grain orientation, which is an advantage since crystal orientation in the seed is at random. For the process, the laser pulse has to be fed through a window in the deposition chamber onto the growing layer. In this way the pulses can be applied without stopping deposition. For a-Si deposition we use electron beam evaporation which has first the advantage of high deposition rate, at least an order of magnitude higher than for PECVD, and secondly the advantage that deposition is directed so that no deposition occurs at the laser window. Doping is achieved by co-deposition of boron or phosphorus. In our device we can deposit and laser irradiate substrates of up to 10x10 cm². The single laser spot has a size of 6x6 mm² with top hat profile. To cover the whole substrate area the laser spot is scanned over the substrate by a scanning mirror placed outside the deposition chamber. In order to avoid cracks in the glass substrate heating to about 600°C helps. Upscaling the system to m² surely is a challenge but not outside the technical possibilities. If properly optimized, about 10 laser pulses are needed at each position during absorber deposition to prepare a 2 μm thick epitaxial film. This makes sense only if the laser is fed into the deposition chamber and is applied without braking deposition, as we do it in our lab scale equipment.

In the epitaxial layer prepared by layered laser crystallization the number of extended defects like dislocations is much lower as compared to solid state epitaxy. This is because the mobility of crystallizing atoms is much higher in the melt than in a-Si so that correct placement is easier. The highest efficiencies achieved in solar cells prepared using the method were 4.8% at an open circuit voltage of 517 mV (Andrä et al., 2005b; Andrä et al., 2007). These values were measured on cells without any light trapping.

3.4.4 Liquid or solid phase epitaxy by diode laser irradiation

The layered laser crystallization method described in the last section has the drawback that up-scaling into the industrial scale is not so easy. This is due to the fact, that the laser beam has to be fed into the deposition chamber and several pulses have to be applied at each position. That was the motivation for us to look for a method in which the complete absorber thickness is deposited in the amorphous state on top of the seed and to apply a single laser treatment to epitaxially crystallize the whole system in one run after deposition outside the deposition chamber.

The most obvious way to achieve epitaxy is via the liquid phase similar to layered laser crystallization. The main difference is that the whole amorphous absorber precursor layer is melted in one step down to the seed, so that epitaxial solidification is to occur after irradiation. It is a challenge to melt about 1 μm of a-Si without completely melt the about 200 nm thin c-Si seed beneath which would hamper any epitaxy. To crystallize a layer system more than 1 μm thick, a short pulse laser is useless. To get the required energy into the system the pulse fluence would have to be so large that ablation would occur at the surface. Moreover, the cooling rate of the melt after a short laser pulse would be so high, that nucleation is expected to occur in a surface near region before the epitaxial solidification front reaches the surface. Therefore we decided to use a scanned cw diode laser for this purpose with irradiation times in the ms range. In this case the cooling rate is low enough so that the melt stays long enough in a slightly undercooled state with low nucleation rate until the epitaxial solidification front reaches the surface. We succeeded in epitaxially crystallizing 500 nm in one run. However, forming of cracks is an issue. Moreover, due to the strong diffusion in the melt which intermixes any pre-existing doping profile, absorber and emitter cannot be crystallized in one step.

An alternative is solid phase epitaxy in which the amorphous layer is heated by the laser to a temperature of about 1100°C, below the melting point of a-Si. At such high temperature the solid phase epitaxial speed was determined to several 100 nm/s high so that epitaxy of 1 μm should be complete within several seconds.

4. Post-crystallization treatment

4.1 Emitter preparation

The emitter of the final solar cell can be prepared in different ways. One is to include emitter doping into the deposition sequence of the layer system so that no additional emitter preparation step is needed. This way has been chosen in the CSG process and in layered laser crystallization. It cannot be applied in case of liquid phase epitaxy of the whole layer stack (Sect. 3.4.4) since during melting for several ms, diffusion in the liquid state would intermix any dopand profile introduced during deposition. In this case, phosphorus doping of a boron doped absorber as in conventional wafer cells can be performed. The only difference is that the doping profile has to be much shallower. Another variant is to use amorphous heteroemitters. IMEC has found that this is the best emitter for their thin film solar cells prepared by the high temperature route (Gordon et al., 2007).

4.2 RTA and hydrogen passivation

To improve the solar cell performance some post-crystallization treatment is required. One point is dopand activation, the other defect passivation.

In order that dopand atoms like boron or phosphorus really lead to a free carrier concentration higher than the intrinsic one, it is necessary that the dopand atoms are included substitutionally in the lattice, i.e. that they rest on regular lattice positions replacing a silicon atom. If they are included interstitially, resting not on regular lattice positions, they are useless. If the silicon lattice forms from the melt the mobility of atoms is high enough so that the dopand atoms can occupy lattice positions. In this case no additional means are needed to make them active. This is not so in case of solid state crystallization. There most of the dopand atoms are included interstitially so that they are inactive. To let them replace silicon atoms substitutionally an additional heat treatment is needed, which is realized by a rapid thermal annealing (RTA) step. In the CSG process, for example, the whole system is heated to about 900°C for 2 min to achieve dopand activation (Keevers et al., 2007). It has been a lot of speculation if this RTA step also improves the grain structure by reducing the number of extended defects. This seems not to be the case (Brazil & Green, 2010).

In any case a hydrogen passivation step has to follow, in which different types of defects e.g. dislocations and grain boundaries, are passivated. Usually, a remote hydrogen plasma is applied to the layer system for 10 to 30 min at about 500°C. Crucial is that during cooling down at the end of the process the plasma has to be applied for some time. A lot of optimization work has been devoted to this passivation step (Rau et al., 2006), which easily can improve the open circuit voltage of the cell by 200 mV.

5. Kinetics of phase transformation

In Sect. 5 the basics of phase transformation relevant for silicon thin film crystallization, both from the melt and in the solid state are summarized (Falk & Andrä, 2006). The Section divides in the propagation of already present phase boundaries and in nucleation, including non-stationary nucleation. Kinetics of aluminum induced crystallization has already been reviewed (Pihan et al., 2007) and is not treated in the following. The facts presented in this section are the background for any successful crystallization of amorphous silicon, in the furnace or by laser irradiation. Quantitative values following from the equations depend on the material parameters of the system involved. These are rather well known for crystalline and for liquid silicon, mostly in the whole range of temperature involved in the processes. This is not the case for amorphous silicon, the properties of which strongly depend on the preparation conditions. They may appreciably differ for hydrogenated a-Si prepared by PECVD and hydrogen free a-Si deposited by electron beam evaporation. Therefore, quantitative predictions have to be taken with some care.

5.1 Propagation of phase boundaries

The propagation speed of already present phase boundaries into a metastable phase, i.e. the growth of a crystal into the undercooled melt or into amorphous silicon, can quantitatively be described by the Jackson-Chalmers-Frenkel-Wilson equation

$$v = v_0 e^{-g^*/kT} \left(1 - e^{\Delta\mu / kT} \right) \tag{1}$$

The prefactor $v_0 = a_0\gamma v$ depends on the atomic vibration frequency (Debye frequency) v, the jump distance a_0 of the order of the lattice parameter of silicon and on a geometry factor γ of the order of 1. $\Delta\mu > 0$ is the difference in chemical potential of the phases involved. For the transition from liquid to crystalline $\Delta\mu$ may be approximated by

$$\Delta\mu = \Delta h_c (1 - \frac{T}{T_{mc}}) \qquad (2)$$

where Δh_c is the latent heat per mole for melting and T_{mc} is the equilibrium melting temperature of 1685 K. For the crystallization of amorphous silicon $\Delta\mu$ is given in the literature (Donovan et al., 1983). g^* is an activation energy for the jump of an atom from the parent to the final phase and is related to the self-diffusion coefficient D according to

$$D = \frac{a_0^2}{\gamma} v \, e^{-g^*/kT} \qquad (3)$$

Results for crystallization from the melt and in the solid state are given in Figs. 6 and 7. In the melt the crystallization speed vanishes at the equilibrium melting point T_{mc} to increase to a maximum of about 16 m/s at 200 K undercooling. At even lower temperature the solidification front gets slower due to the increasing influence of the activation energy. At temperatures above the melting point the phase front runs into the crystal, i.e. the crystal melts and the speed changes sign. In Fig. 6. also the melting speed of amorphous silicon is shown (with opposite sign as compared to c-Si). Melting of a-Si starts at T_{ma}, which, depending on the deposition conditions of a-Si, is 200 to 300 K lower than the melting point of c-Si.

The crystallization speed in amorphous silicon shown in Fig. 7 increases with temperature, and reaches about 1 mm/s near the melting point of a-Si. At 600°C the speed is only about 0.2 nm/s which well correlates with the results obtained in furnace solid phase epitaxy (Sect. 3.4.2).

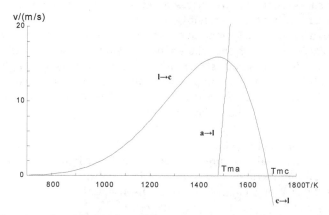

Fig. 6. Speed of the phase boundaries liquid-crystalline (lc) and amorphous liquid (al) for crystalline solidification form the melt and melting of a-Si, respectively.

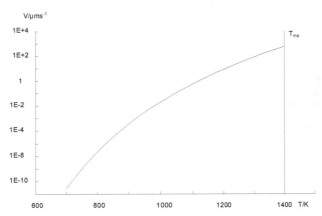

Fig. 7. Speed of a crystallization front in amorphous silicon as depending on temperature

5.2 Stationary nucleation rate

Classical nucleation theory gives the nucleation rate J, i.e. the number of nuclei appearing in a metastable phase per volume and time interval at given temperature. The value applies after some induction time (Sect. 5.3) and as long as not too much of the parent phase is consumed.

$$J = (36\pi)^{1/2} \gamma v \frac{j_c^{2/3}}{V_m} \sqrt{\frac{\Delta G_c}{3 j_c^2 \pi k T}} \, e^{\frac{\Delta G_c + g^*}{kT}} \tag{4}$$

In this formula V_m is the atomic volume and j_c and ΔG_c are the number of atoms in and the free energy of a critical nucleus of the new phase in the matrix of the parent phase, respectively. These are given by

$$j_c = \frac{32\pi}{3} V_m^2 \frac{\sigma^2}{\Delta \mu^2} \tag{5}$$

$$\Delta G_c = \frac{16\pi}{9} V_m^2 \frac{\sigma^3}{\Delta \mu^2} = \frac{1}{2} j_c \Delta \mu \tag{6}$$

σ is the interface energy between both the phases, which, however, is hard to determine independently of nucleation phenomena, and, in addition, may depend on temperature. Moreover, σ strongly influences the nucleation rate since via Eqs. 5&6 it enters Eq. 4 in the third power within the exponential. For crystallization in an undercooled silicon melt the stationary nucleation rate is plotted in Fig. 8 for a temperature dependent interfacial energy according to σ = (43,4+0.249 T/K) mJ/m² (Ujihara et al., 2001). Down to about 300 K below the equilibrium melting point the nucleation rate is very low to change within 100 K of further cooling by 35 orders of magnitude. Below 1200 K the nucleation rate gets rather flat at a value of 10^{35} m⁻³s⁻¹ = 0.1 nm⁻³ns⁻¹. The stationary nucleation rate of crystallization in amorphous silicon is plotted in Fig. 9. There the values increase by 16 orders of magnitude when temperature is increased from 600 K to 1200 K. The nucleation rate then flattens off at 10^{17} m⁻³s⁻¹ = 0.1 μm⁻³s⁻¹ up to the melting point of a-Si of 1400 K.

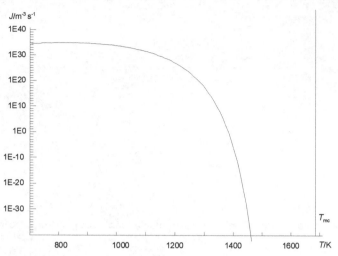

Fig. 8. Stationary nucleation rate for crystallization in an undercooled silicon melt

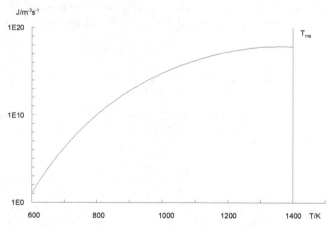

Fig. 9. Stationary nucleation rate for crystallization of amorphous silicon

5.3 Non-stationary nucleation

When the temperature of a system is changed abruptly from a value where the parent phase is absolutely stable and there are no nuclei present to another temperature where it gets metastable, then a population of nuclei evolves. Finally, a stationary distribution of nuclei emerges which leads to the stationary nucleation rate of Eq. 4. The master equation for the population of nuclei can be solved numerically. By some approximations a closed form for non-stationary nucleation rate has been derived (Kashchiev, 1969), which leads to the stationary value after some time lag of nucleation, which is given by

$$\tau = \frac{12}{\pi^2} \frac{kT}{\Delta G_c} \frac{j_c^2}{\beta_c} \tag{7}$$

β_c is the attachment rate of atoms to the critical nucleus given by

$$\beta_c = \gamma g v j_c^{2/3} e^{-g^*/kT} \qquad (8)$$

Here g is an accommodation coefficient of the order of 1. The result for nucleation of c-Si from the melt is shown in Fig. 10. The time lag diverges at the equilibrium melting point and has a minimum of 30 ps around 1350 K. At all relevant temperatures the time lag is so small that it does not play any role in laser crystallization with pulses longer than 1 ns.

This is different for solid phase crystallization of amorphous silicon as shown in Fig. 11. The time lag goes down from 10^{13} s (or 300.000 years) at 600 K to 0.01 s at the melting point of a-Si (1400 K). That means that below 300°C crystallization never occurs whereas in the CSG process of furnace crystallization at 600°C the time lag is in the range of 2 h which does not play a major role when complete crystallization takes 18 h. However, it gives an upper limit for epitaxial growth by furnace annealing as described in Sect. 3.4.2.

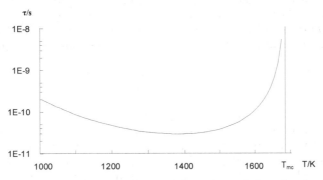

Fig. 10. Time lag of nucleation for crystallization from the melt for a fixed value of interfacial energy σ of 400 mJ/m²

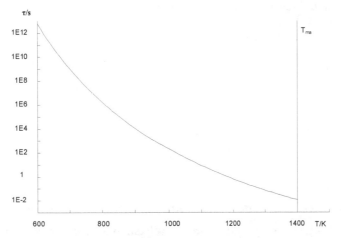

Fig. 11. Time lag of nucleation for crystallization of amorphous silicon

5.4 Complete kinetics of transformation

Stationary nucleation together with the growth of supercritical nuclei according to the Jackson-Chalmers equation leads to a continuous increase of the amount of the new phase on account of the parent phase. When one takes account that during the progress of phase transformation more and more parent phase is consumed and less volume is available for actual transformation, one ends up with the Avrami-Mehl equation (Avrami, 1940) for the volumetric amount of the new phase α

$$\alpha = 1 - e^{t^4/t_c^4} \tag{9}$$

with the characteristic time

$$t_c = \sqrt[4]{\frac{3}{\pi J v^3}} \tag{10}$$

J is the stationary nucleation rate of Eq. 4 and v is the speed of propagation of a phase front according to Jackson-Chalmers Eq. 1. In deriving Eq. 9 the time lag of nucleation τ has been neglected. To include this effect, one simply replaces t by $(t-\tau)$ in Eq. 9 for $t>\tau$. The resulting average grain size when the parent phase has been consumed completely is given by

$$D = 1.037 \sqrt[4]{\frac{v}{J}} \tag{11}$$

So the grains are the larger the higher the Jackson-Chalmers speed and the lower the nucleation rate is, which sounds reasonable. To get large grains from an undercooled melt one should keep the temperature in a range of not too high undercooling, where nucleation rate is low and growth rate is high (Figs. 6 and 8). Fig. 12 shows the expected final grain size in solid phase crystallization of amorphous silicon. It shows that in the CSG process at about 600°C (see Sect. 3.) grains of several μm are to be expected, which is in accordance with experiments. By increasing the crystallization temperature one cannot change the grain size appreciably. Lowering the temperature would lead to a rather high time needed for crystallization due to higher time lag of nucleation (Fig. 11), lower nucleation rate (Fig. 9), and lower growth rate (Fig. 7).

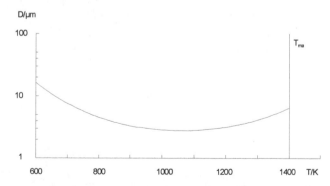

Fig. 12. Average grain size after solid phase crystallization of amorphous silicon as depending on temperature

6. Conclusion

Multi- and polycrystalline silicon thin film solar cells receive growing interest worldwide. Presently, the maximum efficiency reached by these types of cells is 10.4%. Different cell concepts and preparation methods are under investigation and no clear favourite way is identified up to now. The concepts differ in the resulting grain structure, i.e. size and quality, but also in the preparation technologies used and the processing time needed. Today it is not clear which of the methods will succeed in industrial production. In all the methods, pin holes in the films are an issue since they lead to shunting of the final cells. Another issue is dopand deployment, particularly along grain boundaries. This also may lead to shunting, which today limits the open circuit voltage to slightly above 500 mV. A further point is that TCO cannot easily be used as a front contact in superstrate cells since it hardly withstands the temperatures needed for crystallization. Usually a highly doped silicon layer is used instead, which, however, has somewhat low transparency.

Very important for thin film crystalline solar cell is a perfect light management so that about 2 μm of silicon is enough to absorb the solar spectrum. This can be achieved either by structured substrates or by texturing the surface. In the first case, the irregular substrate surface should not influence the crystallization behaviour. In the second case, the rough surface should not increase surface recombination. Generally, passivation of defects and of the surface is a crucial preparation step.

Concerning the theoretical description of the processes involved in crystallization, the basic equations are well understood. However, there are some issues with the material parameters involved, which, particularly for amorphous silicon, strongly depend on deposition conditions and therefore need to be determined individually. But even if numerical predictions may not completely coincide with experiments due to inadequate numerical values of the materials parameters, general trends can reliably be predicted.

All the mentioned issues need further investigation. Careful study of these topics is expected to lead to full exploitation the potential of the material. Multicrystalline thin film cells with a ratio of grain size over film thickness similar to multicrystalline wafer cells should deliver, if prepared correctly, comparable efficiencies. Therefore we expect the poly- and multicrystalline silicon thin film solar cells to gain increasing significance and may replace microcrystalline silicon cells. Multicrystalline silicon also can act as one partner in tandem cells which would further increase the efficiency.

7. Acknowledgment

This work partly was funded by the European Commission under contract 213303 (HIGH-EF), and by the German state of Thuringia via Thüringer Aufbaubank under contract 2008 FE 9160 (SolLUX). We would like to thank J. Lábár and G. Sáfrán (MFA Budapest) for TEM investigations.

8. References

Amkreutz, M.; Müller, J.; Schmidt, M.; Haschke, J.; Hänel, T. & Schulze, T.F. (2009). Optical and electrical properties of electron beam crystallized thin film silicon solar cells on glass substrates. *Proc. 24th Europ. Photovoltaic Solar Energy Conf.*, pp. 2506-2509

Andrä, G.; Bergmann, J.; Falk, F.; Ose, E. & Stafast, H. (1998). Laser Induced Crystallization of Amorphous Silicon Films on Glass for Thin Film Solar Cells. *Physica status solidi (a)*, Vol. 166, pp. 629-634

Andrä, G.; Bergmann, J.; Falk, F. & Ose, E. (2000). Preparation of single crystalline regions in amorphous silicon layers on glass by Ar+ laser irradiation. *Applied Surface Science*, Vol. 154-155, pp. 123-129

Andrä, G.; Bergmann, J. & Falk, F. (2005a). Laser crystallized multicrystalline silicon thin films on glass. *Thin Solid Films*, Vol. 487, pp. 77-80

Andrä, G.; Bergmann, J.; Bochmann, A.; Falk, F.; Dauwe, S. & Kieliba, T. (2005b). Characterization and simulation of multicrystalline LLC-Si thin film solar cells. *Proc. 20th Europ. Photovoltaic Solar Energy Conf.*, pp. 1171-1174

Andrä, G.; Bochmann, A.; Falk, F.; Gawlik, A.; Ose, E. & Plentz J. (2006). Diode laser crystallized multicrystalline silicon thin film solar cells on glass. *Proc. 21st Europ. Photovoltaic Solar Energy Conf.*, pp. 972-975

Andrä, G.; Plentz, J.; Gawlik, A.; Ose, E.; Falk, F. & Lauer, K. (2007). Advances in multi-crystalline LLC-Si thin film solar cells. *Proc. 22nd Europ. Photovoltaic Solar Energy Conf.*, pp. 1967-1970

Andrä, G.; Gimpel, T.; Gawlik, A.; Ose, E.; Bochmann, A.; Christiansen, S.; Sáfrán, G.; Lábár, J.L. & Falk, F. (2008a). Epitaxial Growth of Silicon Thin Films for Solar Cells. *Proc. 23rd Europ. Photovoltaic Solar Energy Conf.* 2008, pp. 2194-2198

Andrä, G.; Lehmann, C.; Plentz, J.; Gawlik, A.; Ose, E. & Falk, F. (2008b). Varying the Layer Structure in Multicrystalline LLC-Silicon Thin-Film Solar Cells. *Proc. 33rd IEEE Photovoltaic Specialists Conf.*, pp. 457-462

Andrä, G.; Bergmann, J.; Gawlik, A.; Höger, I.; Schmidt, T.; Falk, F.; Burghardt, B. & Eberhardt, G. (2010). Laser Induced Crystallization Processes for Multicrystalline Silicon Thin Film Solar Cells. *Proc. 25th Europ. Photovoltaic Solar Energy Conf.* 2010, pp. 3538-3542

Antesberger, T.; Jaeger, C.; Scholz, M. & Stutzmann, M. (2007). Structural and electronic properties of ultrathin polycrystalline Si layers on glass prepared by aluminum-induced layer exchange. *Applied Physics Lett.*, Vol. 91, pp. 201909

Avrami, M. (1940), Kinetics of Phase Change. II Transformation-Time Relations for Random Distribution of Nuclei, *J. Chemical Physics*, Vol. 8, pp. 212-224

Beaucarne, G.; Bourdais, S. ; Slaoui, A. & Poortmans, J. (2004). Thin-film polycrystalline Si solar cells on foreign substrates : film formation at intermediate temperatures (700-1300°C). *Applied Physics A*, Vol. 79, pp. 469-480

Beaucarne, G.; Gordon, I.; van Gestel, D.; Carnel, L. & Poortmans, J. (2006). Thin-film poly-crystalline silicon solar cells: An emerging photovoltaic technology. *Proc. 21st European Photovoltaic Solar Energy Conference*, pp. 721-725

Beaucarne, G. (2007). Silicon thin-film solar cells. *Advances in OptoElectronics*, Vol. 2007, Article ID 36970

Brazil, I. & Green, M.A. (2010). Investigating polysilicon thin film structural changes during rapid thermal annealing of a thin film crystalline silicon on glass solar cell. *J. Materials Sci.: Materials in Electronics*, Vol. 21, pp. 994-999

Brendel, R. (2001). Review of layer transfer process for crystalline thin-film silicon solar cells. *Japanese J. Applied Physics*, Vol. 40, pp. 4431-4439

Brendel, R.; Feldrapp, K.; Horbelt, R. & Auer, R. (2003). 15.4%-efficient and 25 µm-thin crystalline Si solar cell from layer transfer using porous silicon. *Physica status solidi (a)*, Vol. 197, pp. 497-501

Christiansen, S.; Nerding, M.; Eder, C.; Andrae, G.; Falk, F.; Bergmann, J.; Ose, E. & Strunk H.P. (2000). Defect population and electrical properties of Ar+-laser crystallized polycrystalline silicon thin films. *Materials Res. Soc. Symp. Proc.*, Vol. 621, art. Q7.5.1

Colinge, J.P.; Demoulin, E.; Bensahel, D. & Auvert, G. (1982). Use of selective annealing for growing very large grain silicon on insulator films. *Applied Physics Lett.*, Vol. 41, pp. 346-347

Dogan, P.; Rudigier, E.; Fenske, F.; Lee, K.Y.; Gorka, B.; Rau, B.; Conrad, E. & Gall, S. (2008). Structural and electrical properties of epitaxial Si layers prepared by e-beam evaporation. *Thin Solid Films*, Vol. 516, pp. 6989-6993

Donovan, E.P.; Spaepen, F.; Turnbull, D.; Poate, J.M. & Jacobson, D.C. (1983). Heat of crystallization and melting point of amorphous silicon. *Applied Physics Lett.*, Vol. 42, pp. 698-700

Egan, R.; Keevers, M.; Schubert, U.; Young, T.; Evans, R.; Partlin, S.; Wolf, M.; Schneider, J.; Hogg, D.; Eggleston, B.; Green, M.; Falk, F.; Gawlik, A.; Andrä, G.; Werner, M.; Hagendorf, C.; Dogan, P.; Sontheimer, T. & Gall, S. (2009). CSG minimodules using electron-beam evaporated silicon. *Proc. 24th Europ. Photovoltaic Solar Energy Conf. 2009*, pp. 2279-2285

Falk, F. & Andrä, G. (2006), Laser crystallization – a way to produce crystalline silicon films on glass or polymer substrates. *J. Crystal Growth*, Vol. 287, pp. 397-401

Fogarassy, E; de Unamuno, S.; Legagneux, P.; Plais, F.; Pribat, D.; Godard, B. & Stehle, M. (1999). Surface melt dynamics and super lateral growth regime in long pulse duration excimer laser crystallization of amorphous Si films. *Thin Solid Films*, Vol. 337 pp. 143-147

Fuhs, W.; Gall, S.; Rau, B.; Schmidt, M. & Schneider, J. (2004). A novel route to a poly-crystalline silicon thin-film solar cell. *Solar Energy*, Vol. 77, pp. 961-968

Gall, S.; Schneider, J.; Klein, J.; Hübener, K.; Muske, M.; Rau, B.; Conrad, E.; Sieber, J.; Petter, K.; Lips, K.; Stöger-Pollach, M.; Schattschneider, P. & Fuhs, W. (2006). Large-grained polycrystalline silicon on glass for thin-film solar cells. *Thin Solid Films*, Vol. 511, pp. 7-14

Gat, A.; Gerzberg, L.; Gibbons, J.F.; Magee, T.J.; Peng, J. & Hong, J.D. (1978). cw laser anneal of polycrystalline silicon: Crystalline structure, electrical properties. *Applied Physics Lett.*, Vol. 33, pp. 775-778

Gordon, I.; Carnel, L.; Van Gestel, D.; Beaucarne, G. & Poortmans, J. (2007). 8% efficient thin-film polycrystalline-silicon solar cells based on aluminum-induced crystallization and thermal CVD. *Progress in Photovoltaics*, Vol. 15, pp. 575-586

Green, M.A.; Basore, P.A.; Chang, N.; Clugsto, D.; Egan, R.; Evans, R.; Hogg, D.; Jarnason, S.; Keevers, M.; Lasswell, P.; O'Sullivan, J.; Schubert, U., Turner, A.; Wenham, S.R.; & Young, T. (2004). Crystalline silicon on glass (CSG) thin-film solar cell modules. *Solar energy*, Vol. 77, pp. 857-863

Green, M.A. (2007). Thin-film solar cells: review of materials, technologies and commercial status. *J. Materials Science: Materials in Electronics*, Vol. 18, pp. S15-S19

Green, M.A.; Eemery, K.; Hishikawa, Y. & Warta, W. (2011). Solar cell efficiency tables (version 37). *Progress in Photovoltaics*, Vol. 19, pp. 84-92

Gromball, F.; Heemier, J.; Linke, N.; Burchert, M. & Müller, J. (2004). High rate deposition and in situ doping of silicon films for solar cells on glass. *Solar Energy Materials and Solar Cells*, Vol. 84, pp. 71-82

Hatano, M.; Moon, S; Lee, M.; Grigoropoulos, C.P. & Suzuki, K. (2001). Excimer laser-induced melting and resolidification dynamics of silicon thin films. *J. Korean Physical Society*, Vol. 39, pp. S419-S424

Jackson, K.A. & Chalmers, B. (1956), Kinetics of Solidification. *Canadian J. Physics*, Vol. 34, pp. 473-490

Kunz, T.; Burkert, I.; Gawehns, N. & Auer, R. (2008). Crystalline silicon thin-film solar cells on graphite or SiC-ceramic substrates. *Proc. 23rd Europ. Photovoltaic Solar Energy Conf.* (2008), pp. 2202-2204

Kashchiev, D. (1969). Solution of the non-steady state problem in nucleation kinetics. *Surface Science*, Vol. 14, pp. 209-220

Keevers, M.J.; Young, T.L.; Schubert, U. & Green, M.A. (2007). 10% efficient CSG mini-modules, *Proc. 22nd Europ. Photovoltaic Solar Energy Conference*, pp. 1783-1790

Kodera, H. (1963). Diffusion coefficients of impurities in silicon melt. *Japanese J. Applied Physics*, Vol. 2, pp. 212-219

Kuo, C.-C. (2009) Fabrication of large-grain polycrystalline silicon for solar cells. *Laser Physics*, Vol. 19, pp. 143-147

Lichtenstein, N.; Baettig, R.; Brunner, R.; Müller, J.; Valk, B.; Gawlik, A.; Bergmann, J. & Falk, F. (2010). Scalable, High Power Line Focus Diode Laser for Crystallizing of Silicon Thin Films. *Physics Procedia*, Vol. 5, pp. 109-117

Mariucci, L; Pecora, A; Fortunato, G.; Spinella, C. & Bongiorno, C. (2003). Crystallization mechanisms in laser irradiated thin amorphous silicon films. *Thin Solid Films*, Vol. 427, pp. 91-95

Michaud, J.F.; Rogel, R.; Mohammed-Brahim, T. & Sarret, M. (2006). Cw argon ion laser crystallization of silicon films: Structural properties. *J. Non-Crystalline Solids*, Vol. 352, pp. 998-1002

Pihan, E.; Slaoui, A. & Maurice, C. (2007). Growth kinetics and crystallographic properties of polysilicon thin films formed by aluminium-induced crystallization, *J. Crystal Growth*, Vol. 305, pp. 88-98

Rau, B.; Sieber, J.; Schneider, J.; Muske, M.; Stöger-Pollach, M.; Schattschneider, P.; Gall, S. & Fuhs, W. (2004). Low-temperature Si epitaxy on large-grained polycrystalline seed layers by electron-cyclotron resonance chemical vapor deposition. *J. Crystal Growth*, Vol. 270, pp. 396-401

Rau, B.; Conrad, E. & Gall, S. (2006). Influence of post-deposition treatment of absorber layers on poly-Si thin-film solar cells on glass grown by ECRCVD. *Proc. 21st Europ. Photovoltaic Solar Energy Conf.*, pp. 1418-1421

Reuter, M.; Brendle, W.; Tobail, O. & Werner, J.H. (2009). 50 µm thin solar cells with 17.0% efficiency. *Solar Energy Materials and Solar Cells*, Vol. 93, pp. 704-706

Sarikov, A.; Schneider, J.; Klein, J.; Muske, M. & Gall, S. (2006). Theoretical study of the initial stage of the aluminium-induced layer-exchange process. *J. Crystal Growth*, Vol. 287, pp. 442-445

Schneider, J.; Sarikov, A.; Klein, J.; Muske, M.; Sieber, J.; Quinn, T.; Reehal, H.S.; Gall, S. & Fuhs, W. (2006a). A simple model explaining the preferential (100) orientation of silicon thin films made by aluminum-induced layer exchange. *J. Crystal Growth*, Vol. 287, pp. 423-427

Schneider, J.; Schneider, A.; Sarikov, A.; Klein, J.; Muske, M.; Gall, S. & Fuhs, W. (2006b). Aluminum-induced crystallization: Nucleation and growth process. *J. Non-Crystalline Solids*, Vol. 352, pp. 972-97

Schneider, J.; Dore, J.; Christiansen, S.; Falk, F.; Lichtenstein, N.; Valk, B.; Lewandowska, R.; Slaoui, A.; Maeder, X.; Lábár, J.; Sáfrán, G.; Werner, M.; Naumann, V. & Hagendorf, C. (2010). Solar Cells from Crystalline Silicon on Glass Made by Laser Crystallised Seed Layers and Subsequent Solid Phase Epitaxy. *Proc. 25th Europ. Photovoltaic Solar Energy Conf. 2010*, pp. 3573-3676

Sontheimer, T.; Dogan, P.; Becker, C.; Gall, S.; Rech, B.; Schubert, U.; Young, T.; Partlin, S.; Keevers, M. & Egan, R.J. (2009). 6.7% efficient poly-Si thin film mini-modules by high-rate electron-beam evaporation. *Proc. 24th Europ. Photovoltaic Solar Energy Conf. 2009*, pp. 2478-2481

Straub, A.; Inns, D.; Terry, M.L.; Huang, Y.; Widenborg, P.I. & Aberle A.G. (2005). Optimisation of low-temperature silicon epitaxy on seeded glass substrates by ion-assisted deposition. *J. Crystal Growth*, Vol. 280, pp. 385-400

Teplin, C.W.; Branz, H.M.; Jones, K.M.; Romero, M.J.; Stradins, P. & Gall, S. (2007). Hot-wire chemical vapor deposition epitaxy on polycrystalline silicon seeds on glass. *Materials Research Society Symposium Proc.*, Vol. 989-A06-16: Amorphous and Poly-crystalline Thin-Film Silicon Science and Technology 2007, pp. 133-137

Ujihara, T.; Sazaki, G., Fujiwara, K.; Usami, N. & Nakajima, K. (2001). Physical model for the evaluation of solid-liquid interfacial tension in silicon. *J. Applied Physics*, Vol. 90 (2001), pp. 750-755

Van Gestel, D.; Gordon, I.; Bender, H.; Saurel, D.; Vanacken, J.; Beaucarne, G. & Poortmans, J. (2009). Intragrain defects in polycrystalline silicon layers grown by aluminum-induced crystallization and epitaxy for thin-film solar cells. *J. Applied Physics*, Vol. 105, p. 114507

Wang, Z.M. ; Wang, J.Y.; Jeurgens, L.P.H. & Mittemeijer, E.J. (2008). Thermodynamics and mechanism of metal-induced crystallization in immiscible alloy systems: Experiments and calculations on Al/a-Ge and Al/a-Si bilayers. *Physical Review B*, Vol. 77, pp. 045424

Werner, J.H.; Dassow, R.; Rinke, T.J.; Köhler, J.R. & Bergmann, R.B. (2001). From polycrystalline to single crystalline silicon on glass. *Thin Solid Films*, Vol. 383, pp. 95-
101

7

TCO-Si Based Heterojunction Photovoltaic Devices

Z.Q. Ma[1] and B. He[2]

[1]SHU-Solar E PV Laboratory, Department of Physics, Shanghai University, Shanghai
[2]Department of Applied Physics, Donghua University, Shanghai
P. R. China

1. Introduction

It is a common viewpoint that the adscription of the PV research and industry in future has to be the lower cost and higher efficiency. However, those monocrystal as well as multi-crystalline silicon wafer require very expensive processing techniques to produce low defect concentrations, and they are made by complicated wet chemical treatment, high-temperature furnace steps, and time-cost metallization. Thus, a high PV module cost exists for the first-generation technology. Recently, a strong motivation in R&D roadmap of PV cells has been put forward in thin film materials and heterojunction device fields. A large variety of possible and viable methods to manufacture low-cost solar cells are being investigated. Among these strategies, transparent conductive oxides (TCOs) and polycrystalline silicon thin films are promising for application of PV and challenging to develop cheap TCOs and TCO/c-Si heterojunction cells.

Converting solar energy into electricity provides a much-needed solution to the energy crisis in the world is facing today. Solar cells (SC) fabricated on the basis of semiconductor–insulator– semiconductor (SIS) structures are very promising because it is not necessary to obtain a p-n junction and the separation of the charge carriers generated by the solar radiation is realized by the electrical field at the insulator–semiconductor interface. Such SIS structures are obtained by the deposition of thin films of TCO on the oxidized semiconductor surface. One of the main advantages of SIS based SC is the elimination of high temperature diffusion process from the technological chain, the maximum temperature at the SIS structure fabrication by PVD/CVD being not higher than 450 °C. Besides that, the superficial layer of silicon wafer, where the electrical field is localized, is not affected by the impurity diffusion. The TCO films with the band gap in the order of 2.5-4.5 eV are transparent in the whole region of solar spectrum, especially in the blue and ultraviolet regions, which increase the photo response in comparison with the traditional SC. The TCO layer assists the collection of charge carriers and at the same time is an antireflection coating. The most utilized TCO layers are SnO_2, In_2O_3 and their mixture ITO, as well as zinc oxide (ZnO). The efficiency of these kinds of devices can reach the value of more than 10% (Koida et al., 2009).

Transparent conducting oxides (TCOs), such as ZnO, Al-doped ZnO or ITO (SnO_2:In_2O_3), are an increasingly significant component in photovoltaic (PV) devices, where they act as electrodes, structural templates, and diffusion barriers, and their work function are dominant to the open-circuit voltage. The desirable characteristics of TCO materials that are common to all PV technologies are similar to the requirements for TCOs for flat-panel display applications and include high optical transmission across a wide spectrum and low resistivity. Additionally, TCOs for terrestrial PV applications must be used as low-cost materials, and some may be required in the device-technology specific properties. The fundamentals of TCOs and the matrix of TCO properties and processing as they apply to current and future PV technologies were discussed.

As an example, the In_2O_3:SnO_2(ITO) transparent conducting oxides thin film was successfully used for the novel ultraviolet response enhanced PV cell with silicon-based SINP configuration. The realization of ultraviolet response enhancement in PV cells through the structure of ITO/SiO_2/np-Silicon frame (named as SINP), which was fabricated by the state of the art processing, have been elucidated in the chapter. The fabrication process consists of thermal diffusion of phosphorus element into p-type texturized crystal Si wafer, thermal deposition of an ultra-thin silicon dioxide layer (15-20Å) at low temperature, and subsequent deposition of thick In_2O_3:SnO_2 (ITO) layer by RF sputtering. The structure, morphology, optical and electric properties of the ITO film were characterized by XRD, SEM, UV-VIS spectrophotometer and Hall effects measurement, respectively.

The results showed that ITO film possesses high quality in terms of antireflection and electrode functions. The device parameters derived from current-voltage (I-V) relationship under different conditions, spectral response and responsivity of the ultraviolet photoelectric cell with SINP configuration were analyzed in detail. We found that the main feature of our PV cell is the enhanced ultraviolet response and optoelectronic conversion. The improved short-circuit current, open-circuit voltage, and filled factor indicate that the device is promising to be developed into an ultraviolet and blue enhanced photovoltaic device in the future.

On the other hand, the novel ITO/AZO/SiO_2/p-Si SIS heterojunction has been fabricated by low temperature thermally grown an ultrathin silicon dioxide and RF sputtering deposition ITO/AZO double films on p-Si texturized substrate. The crystalline structural, optical and electrical properties of the ITO/AZO antireflection films were characterized by XRD, UV-VIS spectrophotometer, four point probes, respectively. The results show that ITO/AZO films have good quality. The electrical junction properties were investigated by I-V measurement, which reveals that the heterojunction shows strong rectifying behavior under a dark condition. The ideality factor and the saturation current of this diode is 2.3 and 1.075×10^{-5}A, respectively. In addition, the values of I_F/I_R (I_F and I_R stand for forward and reverse current, respectively) at 2V is found to be as high as 16.55. It shows fairly good rectifying behavior indicating formation of a diode between AZO and p-Si. High photocurrent is obtained under a reverse bias when the crystalline quality of ITO/AZO double films is good enough to transmit the light into p-Si.

In device physics, the tunneling effect of SIS solar cell has been investigated in our current work, depending on the thickness of the ultra-thin insulator layer, which is potential for the understanding of quantum mechanics in the photovoltaic devices.

2. Review of TCO thin films

2.1 Development of TCOs
2.1.1 Feature of TCO

Most optically transparent and electrically conducting oxides (TCOs) are binary or ternary compounds, containing one or two metallic elements. Their resistivity could be as low as

10^{-5} Ω cm, and their extinction coefficient k in the visible range (VIS) could be lower than 0.0001, owing to their wide optical band gap (Eg) that could be greater than 3 eV. This remarkable combination of conductivity and transparency is usually impossible in intrinsic stoichiometric oxides; however, it is achieved by producing them with a non-stoichiometric composition or by introducing appropriate dopants. Badeker (1907) discovered that thin CdO films possess such characteristics. Later, it was recognized that thin films of ZnO, SnO_2, In_2O_3 and their alloys were also TCOs. Doping these oxides resulted in improved electrical conductivity without degrading their optical transmission. Al doped ZnO (AZO), tin doped In_2O_3, (ITO) and antimony or fluorine doped SnO_2 (ATO and FTO), are among the most utilized TCO thin films in modern technology. In particular, ITO is used extensively in acoustic wave device, electro-optic modulators, flat panel displays, organic light emitting diodes and photovoltaic devices.

The actual and potential applications of TCO thin films include: (1) transparent electrodes for flat panel displays (2) transparent electrodes for photovoltaic cells, (3) low emissivity windows, (4) window defrosters, (5) transparent thin films transistors, (6) light emitting diodes, and (7) semiconductor lasers. As the usefulness of TCO thin films depends on both their optical and electrical properties, both parameters should be considered together with environmental stability, abrasion resistance, electron work function, and compatibility with substrate and other components of a given device, as appropriate for the application. The availability of the raw materials and the economics of the deposition method are also significant factors in choosing the most appropriate TCO material. The selection decision is generally made by maximizing the functioning of the TCO thin film by considering all relevant parameters, and minimizing the expenses. TCO material selection only based on maximizing the conductivity and the transparency can be faulty.

Recently, the scarcity and high price of Indium needed for ITO materials, the most popular TCO, as spurred R&D aimed at finding a substitute. Its electrical resistivity (ρ) should be ~10^{-4} Ω cm or less, with an absorption coefficient (α) smaller than 10^4 cm^{-1} in the near-UV and VIS range, and with an optical band gap >3eV. A 100 nm thick film TCO film with these values for and will have optical transmission (T) 90% and a sheet resistance (R_S) of < 10 $\Omega/\square$. At present, AZO and ZnO:Ga (GZO) semiconductors are promising alternatives to ITO for thin-film transparent electrode applications. The best candidates is AZO, which can have a low resistivity, e.g. on the order of 10^{-4} Ω cm, and its source materials are inexpensive and non-toxic. However, the development of large area, high rate deposition techniques is needed.

Another objective of the recent effort to develop novel TCO materials is to deposit p-type TCO films. Most of the TCO materials are n-type semiconductors, but p-type TCO materials are required for the development of solid lasers, as well as TFT or PV cells. Such p-type TCOs include: ZnO:Mg, ZnO:N, ZnO:In, NiO, NiO:Li, $CuAlO_2$, Cu_2SrO_2, and $CuGaO_2$ thin films. These materials have not yet found a place in actual applications owing to the stability.

Published reviews on TCOs reported exhaustively on the deposition and diagnostic techniques, on film characteristics, and expected applications. The present paper has three objectives: (1) to review the theoretical and experimental efforts to explore novel TCO materials intended to improve the TCO performance, (2) to explain the intrinsic physical limitations that affect the development of an alternative TCO with properties equivalent to those of ITO, and (3) to review the practical and industrial applications of existing TCO thin films.

2.1.2 Multiformity of TCOs

The first realization of a TCO material (CdO, Badeker 1907)) occurred slightly more than a century ago when a thin film of sputter deposited cadmium (Cd) metal underwent incomplete thermal oxidation upon postdeposition heating in air. Later, CdO thin films were achieved by a variety of deposition techniques such as reactive sputtering, spray pyrolysis, activated reactive evaporation, and metal organic vapor phase epitaxy (MOVPE). CdO has a face centered cubic (FCC) crystal structure with a relatively low intrinsic band gap of 2.28 eV. Note that without doping, CdO is an n-type semiconductor. The relatively narrow band gap of CdO and the toxicity of Cd make CdO less desirable and account for receiving somewhat dismal attention in its standard form. However, its low effective carrier mass allows efficiently increasing the band gap of heavily doped samples to as high as 3.35 eV (the high carrier concentration results in a partial filling of a conduction band and consequently, in a blue-shift of the UV absorption edge, known as the Burstein–Moss effect) and gives rise to mobility as high as 607 cm^2/V s in epitaxial CdO films doped with Sn. The high mobility exhibited by doped CdO films is a definite advantage in device applications. Cd-based TCOs such as CdO doped with either indium (In), tin (Sn), fluorine (F), or yttrium (Y), and its ternary compounds such as $CdSnO_3$, Cd_2SnO_4, $CdIn_2O_4$ as well as its other relevant compounds all have good electrical and optical properties. The lowest reported resistivity of Cd-based TCOs is 1.4×10^{-4} Ω cm, which is very good and competitive with other leading candidates. The typical transmittance of Cd-based TCOs in the visible range is 85%–90%. Although the Cd-based TCOs have the desired electrical and optical properties, in addition to low surface recombination velocity, which is very desirable, they face tremendous obstacles in penetrating the market except for some special applications such as CdTe/CdS thin film solar cells due to the high toxicity of Cd. It should be noted that the aforementioned solar cells are regulated and cannot be sold. To circumvent this barrier, the manufacturers lease them for solar power generation instead. Consequently, our attention in this chapter is turned away for discussing this otherwise desirable conducting oxide.

Revelations dating back to about 1960s that indium tin oxide (ITO), a compound of indium oxide (In_2O_3) and tin oxide (SnO_2), exhibits both excellent electrical and optical properties paved the way for extensive studies on this material family. In_2O_3 has a bixbyite-type cubic crystal structure, while SnO_2 has a rutile crystal structure. Both of them are weak n-type semiconductors. Their charge carrier concentration and thus, the electrical conductivity can be strongly increased by extrinsic dopants which is desirable. In_2O_3 is a semiconductor with a band gap of 2.9 eV, a figure which was originally thought to be 3.7 eV. The reported dopants for In_2O_3-based binary TCOs are Sn, Ge, Mo, Ti, Zr, Hf, Nb, Ta, W, Te, and F as well as Zn. The In_2O_3-based TCOs doped with the aforementioned impurities were found to possess very good electrical and optical properties. The smallest laboratory resistivities of Sn-doped In_2O_3 (ITO) are just below 10^{-4} Ω cm, with typical resistivities being about 1×10^{-4} Ω cm. As noted above, despite the nomenclature of Sn-doped In_2O_3 (ITO), this material is really an In_2O_3-rich compound of In_2O_3 and SnO_2. SnO_2 is a semiconductor with a band gap of 3.62 eV at 298 K and is particularly interesting because of its low electrical resistance coupled with its high transparency in the UV–visible region. SnO_2 grown by molecular beam epitaxy (MBE) was found to be unintentionally doped with an electron concentration for different samples in the range of $(0.3–3) \times 10^{17}$ cm^{-3} and a corresponding electron mobility in the range of 20–100 cm^2/V s. Fluorine (F), antimony (Sb), niobium (Nb), and tantalum (Ta) are most commonly used to achieve high n-type conductivity while maintaining high optical transparency.

Much as ITO is the most widely used In_2O_3-based binary TCO, fluorine-doped tin oxide (FTO) is the dominant in SnO_2-based binary TCOs. In comparison to ITO, FTO is less expensive and shows better thermal stability of its electrical properties as well chemical stability in dye-sensitized solar cell (DSSC). FTO is the second widely used TCO material, mainly in solar cells due to its better stability in hydrogen-containing environment and at high temperatures required for device fabrication. The typical value of FTO's average transmittance is about 80%. However, electrical conductivity of FTO is relatively low and it is more difficult to pattern via wet etching as compared to ITO. In short, more efforts are beginning to be expended for TCOs by researchers owing to their above-mentioned uses spurred by their excellent electrical and optical properties in recently popularized devices. Germanium-doped indium oxide, IGO (In_2O_3:Ge), and fluorine-doped indium oxide, IFO (In_2O_3:F), reported by Romeo et al., for example, have resistivities of about $2 \times 10^{-4} \Omega$ cm and optical transmittance of $\geq 85\%$ in the wavelength range of 400–800 nm, which are comparable to their benchmark ITO. Molybdenum-doped indium oxide, IMO (In_2O_3:Mo), was first reported by Meng et al.. Later on, Yamada et al. reported a low resistivity of $1.5 \times 10^{-4} \Omega$ cm and a mobility of 94 cm^2/V s, and Parthiban et al. reported a resistivity of $4 \times 10^{-4} \Omega$ cm, an average transmittance of >83% and a mobility of 149 cm^2/V s for IMO. Zn-doped indium oxide, IZO (In_2O_3:Zn), deposited on plastic substrates showed resistivity of $2.9 \times 10^{-4} \Omega$ cm and optical transmittance of $\geq 85\%$. Suffice it to say that In_2O_3 doped with other impurities have comparable electrical and optical properties to the above-mentioned data as enumerated in many articles.

The small variations existing among these reports could be attributed to the particulars of the deposition techniques and deposition conditions. To improve the electrical and optical properties of In_2O_3 and ITO, their doped varieties such as ITO:Ta and In_2O_3:Cd–Te have been explored as well. For example, compared with ITO, the films of ITO:Ta have improved the electrical and optical properties due to the improved crystallinity, larger grain size, and the lower surface roughness, as well as a larger band gap, which are more pronounced for ITO:Ta achieved at low substrate temperatures. The carrier concentration, mobility, and maximum optical transmittance for ITO:Ta achieved at substrate temperature 400°C are 9.16×10^{20} cm^{-3}, 28.07 cm^2/V s and 91.9% respectively, while the corresponding values for ITO are 9.12×10^{20} cm^{-3}, 26.46 cm^2/V s and 87.9%, respectively. Due to historical reasons, propelled by the above discussed attributes, ITO is the predominant TCO used in optoelectronic devices. Another reason why ITO enjoys such predominance is the ease of its processing. ITO-based transparent electrodes used in LCDs consume the largest amount of indium, about 80% of the total. As reported by Minami and Miyata (January, 2008), about 800 tons of indium was used in Japan in 2007. Because approximately 80%–90% of the indium can be recycled, the real consumption of indium in Japan in 2007 is in the range of 80–160 tons. The total amount of indium reserves in the world is estimated to be only approximately 6000 tons according to the 2007 United States Geological Survey. It is widely believed that indium shortage may occur in the very near future and indium will soon become a strategic resource in every country.

Consequently, search for alternative TCO films comparable to or better than ITO is underway. The report published by NanoMarkets in April 2009 (Indium Tin Oxide and Alternative Transparent Conductor Markets) pointed out that up until 2009 the ITO market was not challenged since the predicted boom in demand for ITO did not happen, partially due to the financial meltdown. The price of indium slightly varied from about US700$/kg in 2005 to US1000$/kg in 2007 and then to US700$/kg in 2009 which is still too expensive for mass production. On the other hand, the market research firm iSupply forecasted in 2008 that the worldwide market for all touch screens employing ITO layers would nearly double, from \$3.4 billion to \$6.4 billion by 2013. Therefore, ITO as the industrial standard TCO is expected to lose its share of the applicable markets rather slowly even when alternatives become available. The report by NanoMarkets is a good guide for both users and manufacturers of TCOs.

In addition to ZnO-based TCOs, it also remarks on other possible solutions such as conductive polymers and/or the so-called and overused concept of nano-engineered materials such as poly (3, 4-ethylenedioxythiophene) well known as PEDOT by both H.C. Starck and Agfa, and carbon nanotube (CNT) coatings, which have the potentials to replace ITO at least in some applications since they can overcome the limitations of TCOs. Turning our attention now to the up and coming alternatives to ITO, ZnO with an electron affinity of 4.35 eV and a direct band gap energy of 3.30 eV is typically an n-type semiconductor material with the residual electron concentration of~10^{17} cm^{-3}. However, the doped ZnO films have been realized with very attractive electrical and optical properties for electrode applications. The dopants that have been used for the ZnO-based binary TCOs are Ga, Al, B, In, Y, Sc, V, Si, Ge, Ti, Zr, Hf, and F. Among the advantages of the ZnO-based TCOs are low cost, abundant material resources, and non-toxicity. At present, ZnO heavily doped with Ga and Al (dubbed GZO and AZO) has been demonstrated to have low resistivity and high transparency in the visible spectral range and, in some cases, even outperform ITO and FTO. The dopant concentration in GZO or AZO is more often in the range of 10^{20}–10^{21} cm^{-3} and although we obtained mobilities near 95 cm^2/V s in our laboratory in GZO typical reported mobility is near or slightly below 50 cm^2/V s. Ionization energies of Al and Ga donors (in the dilute limit which decreases with increased doping) are 53 and 55 meV, respectively, which are slightly lower than that of In (63 meV). Our report of a very low resistivity of~8.5×10^{-5} Ω cm for AZO, and Park et al. reported a resistivity of ~8.1×10^{-5} Ω cm for GZO, both of which are similar to the lowest reported resistivity of~7.7×10^{-5} Ω cm for ITO. The typical transmittance of AZO and GZO is easily 90% or higher, which is comparable to the best value reported for ITO when optimized for transparency alone and far exceeds that of the traditional semi-transparent and thin Ni/Au metal electrodes with transmittance below 70% in the visible range. The high transparency of AZO and GZO originates from the wide band gap nature of ZnO. Low growth temperature of AZO or GZO also intrigued researchers with respect to transparent electrode applications in solar cells. As compared to ITO, ZnO-based TCOs show better thermal stability of resistivity and better chemical stability at higher temperatures, both of which bode well for the optoelectronic devices in which this material would be used. In short, AZO and GZO are the TCOs attracting more attention, if not the most, for replacing ITO. From the cost and availability and environmental points of view, AZO appears to be the best candidate. This conclusion is also bolstered by batch process availability for large-area and large-scale production of AZO.

To a lesser extent, other ZnO-based binary TCOs have also been explored. For readers'convenience, some references are discussed at a glance below. B-doped ZnO has been reported to exhibit a lateral laser-induced photovoltage (LPV), which is expected to make it a candidate for position sensitive photo-detectors. In-doped ZnO prepared by pulsed laser deposition and spray pyrolysis is discussed, respectively. Y-doped ZnO deposited by sol–gel method on silica glass has been reported. The structural, optical and electrical properties of F-doped ZnO formed by the sol–gel process and also listed almost all the relevant activities in the field. For drawing the contrast, we should reiterate that among all the dopants for ZnO-based binary TCOs, Ga and Al are thought to be the best candidates so far. It is also worth nothing that $Zn_{1-x}Mg_xO$ alloy films doped with a donor impurity can also serve as transparent conducting layers in optoelectronic devices. As well known the band gap of wurtzite phase of $Zn_{1-x}Mg_xO$ alloy films could be tuned from 3.37 to 4.05 eV, making conducting $Zn_{1-x}Mg_xO$ films more suitable for ultraviolet (UV) devices. The larger band gap of these conducting layers with high carrier concentration is also desired in the modulation-doped heterostructures designed to increase electron mobility. In this vein, $Zn_{1-x}Mg_xO$ doped with Al has been reported in Refs. The above-mentioned ZnO-based TCOs have relatively large refractive indices as well, in the range of 1.9–2.2, which are comparable to those of ITO and FTO. For comparison, the refractive indices of commercial ITO/glass decrease from 1.9 at wavelength of 400 nm to 1.5 at a wavelength of 800 nm, respectively. The high refractive indices reduce internal reflections and allow employment of textured structures in LEDs to enhance light extraction beyond that made feasible by enhanced transparency alone. The dispersion in published values of the refractive index is attributed to variations in properties of the films prepared by different deposition techniques. For example, amorphous ITO has lower refractive index than textured ITO. It is interesting to note that nanostructures such as nanorods and nanotips as well as controllable surface roughness could enhance light extraction/absorption in LEDs and solar cells, thus improving device performance. Fortunately, such nanostructures can be easily achieved in ZnO by choosing and controlling the growth conditions. One disadvantage of ZnO-based TCOs is that they degrade much faster than ITO and FTO when exposed to damp and hot (DH) environment. The stability of AZO used in thin film $CuInGaSe_2$ (CIGS) solar cells, along with Al-doped $Zn_{1-x}Mg_xO$ alloy, ITO and FTO, by direct exposure to damp heat (DH) at 85°C and 85% relative humidity. The results showed that the DH-induced degradation rates followed the order of AZO and $Zn_{1-x}Mg_xO \gg$ ITO > FTO. The degradation rates of AZO were slower for films of larger thickness which were deposited at higher substrate temperatures during sputter deposition, and underwent dry-out intervals. From the point of view of the initiation and propagation of degrading patterns and regions, the degradation behavior appears similar for all TCOs despite the obvious differences in the degradation rates. The degradation is explained by both hydrolysis of the oxides at some sporadic weak spots followed by swelling and popping of the hydrolyzed spots which are followed by segregation of hydrolyzed regions, and hydrolysis of the oxide–glass interfaces.

In addition to those above-mentioned binary TCOs based on In_2O_3, SnO_2 and ZnO, ternary compounds such as Zn_2SnO_4, $ZnSnO_3$, $Zn_2In_2O_5$, $Zn_3In_2O_6$, $In_4Sn_3O_{12}$, and multicomponent oxides including $(ZnO)_{1-x}(In_2O_3)x$, $(In_2O_3)_x(SnO_2)_{1-x}$, $(ZnO)_{1-x}(SnO_2)_x$ are also the subject of investigation. However, it is relatively difficult to deposit those TCOs with desirable optical and electrical properties due to the complexity of their compositions. Nowadays ITO, FTO and GZO/AZO described in more details above are preferred in practical applications due to the relative ease by which they can be formed. Although it is not within the scope of this article, it has to be pointed out for the sake of completeness that CdO along with In_2O_3 and SnO_2 forms an analogous In_2O_3–SnO_2–CdO alloy system. The averaged resistivity of ITO by different techniques is $\sim 1 \times 10^{-4} \Omega \bullet cm$, which is much higher than that of FTO. For FTO, the typically employed technique is spray pyrolysis which can produce the lowest resistivity of $\sim 3.8 \times 10^{-4}$ $\Omega \bullet cm$. For AZO/GZO, the resistivities listed here are comparable to or slightly higher than ITO but their transmittance is slightly higher than that of ITO. Obviously, AZO and GZO as well as other ZnO-based TCOs are promising to replace ITO for transparent electrode applications in terms of their electrical and optical properties.There are also few reports for some other promising n-type TCOs, which could find some practical applications in the future. They are titanium oxide doped with Ta or Nb, Ga_2O_3 doped with Sn and $12CaO \cdot 7Al_2O_3$ (often denoted $C_{12}A_7$). These new TCOs are currently not capable of competing with ITO/FTO/GZO/AZO in terms of electrical or optical properties. We should also point out that n-type transparent oxides under discussion are used on top of the p-type semiconductors and the vertical conduction between the two relies on tunneling and leakage. The ideal option would be to develop p-type TCOs which are indeed substantially difficult to attain.

3. Crystal chemistry of ITO

Crystalline indium oxide has the bixbyite structure consisting of an 80-atom unit cell with the Ia3 space group and a 1-nm lattice parameter in an arrangement that is based on the stacking of InO_6 coordination groups. The structure is closely related to fluorite, which is a face-centered cubic array of cations with all the tetrahedral interstitial positions occupied with anions. The bixbyite structure is similar to fluorite except that the MO_8 coordination units (oxygen position on the corners of a cube and M located near the center of the cube) of fluorite are replaced with units that have oxygen missing from either the body or the face diagonal. The removal of two oxygen ions from the metal-centered cube to form the InO_6 coordination units of bixbyite forces the displacement of the cation from the center of the cube. In this way, indium is distributed in two nonequivalent sites with one-fourth of the indium atoms positioned at the center of a trigonally distorted oxygen octahedron (diagonally missing O). The remaining three-fourths of the indium atoms are positioned at the center of a more distorted octahedron that forms with the removal of two oxygen atoms from the face of the octahedron. These MO_6 coordination units are stacked such that one-fourth of the oxygen ions are missing from each {100} plane to form the complete bixbyite structure. A minimum in the thin-film resistivity is found in the ITO system when the oxygen partial pressure during deposition is optimized. This is because doping arises from two sources, four-valent tin substituting for three-valent indium in the crystal and the creation of doubly charged oxygen vacancies. This is due to an oxygen-dependent competition between substitutional Sn and Sn in the form of neutral oxide complexes that do not contribute carriers. Amorphous ITO that has been optimized with respect to oxygen content during deposition has a characteristic carrier mobility (40 cm^2/V s) that is only slightly less than that of crystalline films of the same composition. This is in sharp contrast to amorphous covalent semiconductors such as Si, where carrier transport is severely limited by the disorder of the amorphous phase. In semiconducting oxides formed from heavy-metal cations with $(n-1)d^{10}ns^0$ ($n \leq 4$) electronic configurations, it appears that the degenerate band conduction is not band-tail limited.

4. ZnO thin films

Another important oxide used in PV window and display technology applications is doped ZnO, which has been learned to have a thin-film resistivity as low as 2.4 $\times 10^{-4}$ $\Omega \bullet cm$. Although the resistivity of ZnO thin films is not yet as small as the ITO standard, it does offer the significant benefits of low cost relative to In-based systems and high chemical and thermal stability. In the undoped state, zinc oxide is highly resistive because, unlike In-based systems, ZnO native point defects are not efficient donors. However, reasonable impurity doping efficiencies can be achieved through substitutional doping with Al, In, or Ga. Most work to date has focused on Al - doped ZnO, but this dopant requires a high degree of control over the oxygen potential in the sputter gas because of the high reactivity of Al with oxygen. Gallium, however, is less reactive and has a higher equilibrium oxidation potential, which makes it a better choice for ZnO doping applications. Furthermore, the slightly smaller bond length of Ga–O (1.92Å) compared with Zn–O (1.97 Å) also offers the advantage of minimizing the deformation of the ZnO lattice at high substitutional gallium concentrations. The variety of ZnO thin films has been expatiated elsewhere.

5. Electrical conductivity of TCO

TCOs are wide band gap (E_g) semiconducting oxides, with conductivity in the range of 10^2 – 1.2×10^6 (S). The conductivity is due to doping either by oxygen vacancies or by extrinsic dopants. In the absence of doping, these oxides become very good insulators, with the resistivity of $> 10^{10}$ Ω cm. Most of the TCOs are n-type semiconductors. The electrical conductivity of n-type TCO thin films depends on the electron density in the conduction band and on their mobility: $\sigma = \mu\, n\, e$, where μ is the electron mobility, n is its density, and e is the electron charge. The mobility is given by:

$$\mu = e\,\tau\,/\,m^* \tag{1}$$

where τ is the mean time between collisions, and m^* is the effective electron mass. However, as n and τ are negatively correlated, the magnitude of μ is limited. Due to the large energy gap ($E_g > 3$ eV) separating the valence band from the conducting band, the conduction band can not be thermally populated at room temperature (kT~0.03 eV, where k is Boltzmann's constant), hence, stoichiometric crystalline TCOs are good insulators. To explain the TCO characteristics, the various popular mechanisms and several models describing the electron mobility were proposed.

In the case of intrinsic materials, the density of conducting electrons has often been attributed to the presence of unintentionally introduced donor centers, usually identified as metallic interstitials or oxygen vacancies that produced shallow donor or impurity states located close to the conduction band. The excess donor electrons are thermally ionized at room temperature, and move into the host conduction band. However, experiments have been inconclusive as to which of the possible dopants was the predominant donor. Extrinsic dopants have an important role in populating the conduction band, and some of them have been unintentionally introduce. Thus, it has been conjectured in the case of ZnO that interstitial hydrogen, in the H^+ donor state, could be responsible for the presence of carrier electrons. In the case of SnO_2, the important role of interstitial Sn in populating the conducting band, in addition to that of oxygen vacancies, was conclusively supported by first-principle calculations. They showed that Sn interstitials and O vacancies, which dominated the defect structure of SnO_2 due to the multivalence of Sn, explained the natural nonstoichiometry of this material and produced shallow donor levels, turning the material into an intrinsic n-type semiconductor. The electrons released by these defects were not compensated because acceptor-like intrinsic defects consisting of Sn voids and O interstitials did not form spontaneously. Furthermore, the released electrons did not make direct optical transitions in the visible range due to the large gap between the Fermi level and the energy level of the first unoccupied states. Thus, SnO_2 could have a carrier density with minor effects on its transparency.

The conductivity σ is intrinsically limited for two reasons. First, n and τ cannot be independently increased for practical TCOs with relatively high carrier concentrations. At high conducting electron density, carrier transport is limited primarily by ionized impurity scattering, i.e., the Coulomb interactions between electrons and the dopants. Higher doping concentration reduces carrier mobility to a degree that the conductivity is not increased, and it decreases the optical transmission at the near-infrared edge. With increasing dopant concentration, the resistivity reaches a lower limit, and does not decrease beyond it, whereas the optical window becomes narrower. Bellingham were the first to report that the mobility and hence the resistivity of transparent conductive oxides (ITO, SnO_2, ZnO) are limited by ionized impurity scattering for carrier concentrations above 10^{20} cm^{-3}. Ellmer also showed that in ZnO films deposited by various methods, the resistivity and mobility were nearly independent of the deposition method and limited to about 2×10^{-4} Ω cm and 50 cm^2/Vs, respectively. In ITO films, the maximum carrier concentration was about 1.5×10^{21} cm^{-3}, and the same conductivity and mobility limits also held. This phenomenon is a universal property of other semiconductors. Scattering by the ionized dopant atoms that are homogeneously distributed in the semiconductor is only one of the possible effects that reduce the mobility. The all recently developed TCO materials, including doped and undoped binary, ternary, and quaternary compounds, also suffer from the same limitations. Only some exceptional samples had a resistivity of $\leq 1\times10^{-4}$ Ω cm.

In addition to the above mentioned effects that limit the conductivity, high dopant concentration could lead to clustering of the dopant ions, which increases significantly the scattering rate, and it could also produce nonparabolicity of the conduction band, which has to be taken into account for degenerately doped semiconductors with filled conduction bands.

6. Optical properties of TCO

The transmission window of TCOs is defined by two imposed boundaries. One is in the near-UV region determined by the effective band gap Eg, which is blue shifted due to the Burstein–Moss effect. Owing to high electron concentrations involved the absorption edge is shifted to higher photon energies. The sharp absorption edge near the band edge typically corresponds to the direct transition of electrons from the valence band to the conduction band. The other is at the near infrared (NIR) region due to the increase in reflectance caused by the plasma resonance of electron gas in the conduction band. The absorption coefficient (α) is very small within the defined window and consequently transparency is very high. The positions of the two boundaries defining the transmission window are closely related to the carrier concentration. For TCOs, both boundaries defining the transmission window shift to shorter wavelength with the increase of carrier concentration. The blue-shift of the near-UV and near-IR boundaries of the transmission window of GZO as the carrier concentration increased from 2.3×10^{20} cm^{-3} to 10×10^{20} cm^{-3}. The blue-shift of the onset of absorption in the near-UV region is associated with the increase in the carrier concentration blocking the lowest states (filled states) in the conduction band from absorbing the photons. The Burstein–Moss effect owing to high electron concentrations has been widely observed in transmittance spectra of GZO and AZO. A comparable or even larger blue-shift in the transmittance spectra of GZO has been reported with absorption edge at about 300 nm wavelength corresponding to a bang gap of about 4.0 eV. The plasma frequency at which the free carriers are absorbed has a negative correlation with the free carrier concentration.

Consequently, the boundary in the near-IR region also shifts to the shorter wavelength with increase of the free carrier concentration. The shift in the near-IR region is more pronounced than that in the near-UV region. Therefore, the transmission window becomes narrower as the carrier concentration increases. This means that both the conductivity and the transmittance window are interconnected since the conductivity is also related to the carrier concentration as discussed above. Thus, a compromise between material conductivity and transmittance window must be struck, the specifics of which being application dependent. While for LED applications the transparency is needed only in a narrow range around the emission wavelengths, solar cells require high transparency in the whole solar spectral range. Therefore, for photovoltaics, the carrier concentration should be as low as possible for reducing the unwanted free carrier absorption in the IR spectral range, while the carrier mobility should be as high as possible to retain a sufficiently high conductivity. Optical measurements are also commonly employed to gain insight into the film quality. For example, interference fringes found in transmittance curves indicate the highly reflective nature of surfaces and interfaces in addition to the low scattering and absorption losses in the films. The particulars of interferences are related to both the film thickness and the incident wavelength, which can be used to achieve higher transmittance for TCOs. In the case of a low quality TCO, deep level emissions occurring in photoluminescence (PL) spectra along with relatively low transmittance are attributed to the lattice defects such as oxygen vacancies, zinc vacancies, interstitial metal ions, and interstitial oxygen. High-doping concentration-induced defects in crystal lattices causing the creation of electronic defect states in band gap similarly have an adverse effect on transparency. In GZO, as an example, at very high Ga concentrations (10^{20}–10^{21} cm^{-3}), the impurity band merges with the conduction band causing a tail-like state below the conduction band edge of intrinsic ZnO. These tail states are responsible for the low-energy part of PL emission. Therefore, the defects, mainly the oxygen-related ones, in TCOs have to be substantially reduced, if not fully eliminated, through the optimal growth conditions to attain higher transmittance.

7. Application of TCO in solar cells

Solar cells exploit the photovoltaic effect that is the direct conversion of incident light into electricity. Electron–hole pairs generated by solar photons are separated at a space charge region of the two materials with different conduction polarities. Solar cells represent a very promising renewable energy technology because they provide clean energy source (beyond manufacturing) which will reduce our dependence on fossil oil. The principles of operation of solar cells have been widely discussed in detail in the literature and as such will not be repeated here. Rather, the various solar cell technologies will be discussed in the context of conduction oxides. Solar cells can be categorized into bulk devices (mainly single-crystal or large-grain polycrystalline Si), thin film single- and multiple-junction devices, and newly emerged technology which include dye-sensitized cells, organic/polymer cells, high-efficiency multi-junction cells based on III–V semiconductors among others. Crystalline silicon modules based on bulk wafers have been dubbed as the "first-generation" photovoltaic technology. The cost of energy generated by PV modules based on bulk-Si wafers is currently around \$3–\$4/Wp and cost reduction potential seems limited by the price of Si wafers. This cost of energy is still too high for a significant influence on energy production markets. Much of the industry is focused on the most cost efficient technologies in terms of cost per generated power. The two main strategies to bring down the cost of photovoltaic electricity are increasing the efficiency of the cells and decreasing their cost per unit area. Thin film devices (also referred to as second generation of solar cells) consume less material than the bulk-Si cells and, as a result, are less expensive. The market share of the thin film solar cells is continuously growing and has reached some 15% in year 2010, while the other 85% is silicon modules based on bulk wafers. Alternative approaches also focused on reducing energy price are devices based on polymers and dyes as the absorber materials, which include a wide variety of novel concepts. These cells are currently less efficient than the semiconductor-based devices, but are attractive due to simplicity and low cost of fabrication.

TCO are utilized as transparent electrodes in many types of thin film solar cells, such as a-Si thin film solar cells, CdTe thin film solar cells, and CIGS thin film solar cells. It should be mentioned that, for photovoltaic applications, a trade-off between the sheet resistance of a TCO layer and its optical transparency should be made. As mentioned above, to reduce unwanted free carrier absorption in the IR range, the carrier concentration in TCO should be as low as possible, while the carrier mobility should be as high as possible to obtain sufficiently high conductivity. Therefore, achieving TCO films with high carrier mobility is crucial for solar cell applications.

7.1 Si thin film solar cells

In addition to the well-established Si technology and non-toxic nature and abundance of Si, the advantage of thin film silicon solar cells is that they require lower amount of Si as compared to the devices based on bulk wafers and therefore are less expensive. Several different photovoltaic technologies based on Si thin films have been proposed and implemented: hydrogenated amorphous Si (a-Si:H) with quasi-direct band gap of 1.8 eV, hydrogenated microcrystalline Si (μc-Si:H) with indirect band gap of 1.1 eV, their combination (micromorph Si), and polycrystalline Si on glass (PSG) solar cells. The first three technologies rely on TCOs as front/back electrodes. This thin film p–i–n solar cell is fabricated in a so-called superstrate configuration, in which the light enters the active region through a glass substrate. In this case, the fabrication commences from the front of the cell and proceeds to its back.

First, a TCO front contact layer is deposited on a transparent glass substrate, followed by deposition of amorphous/microcrystalline Si, and a TCO/metal back contact layer. Therefore, the TCO front contact must be sufficiently robust to survive all subsequent deposition steps and post-deposition treatments. To obtain high efficiency increasing the path length of incoming light is crucial, which is achieved by light scattering at the interface between Si and TCO layers with different refractive indices, so that light is "trapped" within the Si absorber layer. The light trapping allows reduction of the thickness of the Si absorber layer which paves the way for increased device stability. Therefore, TCO layers used as transparent electrodes in the Si solar cells have a crucial impact on device performance. In addition to high transparency and high electrical conductivity, a TCO layer used as front electrode should ensure efficient scattering of the incoming light into the absorber layer and be chemically stable in hydrogen-containing plasma used for Si deposition, and act as a good nucleation layer for the growth of microcrystalline Si. The bottom TCO layer between Si and a metal contact works as an efficient back reflector as well as a diffusion barrier.

To increase light scattering, surface texturing of the front and back TCO contact layers is commonly used. As discussed above, the TCOs for practical applications are ITO, FTO and

GZO/AZO. For reasons mentioned in the text dealing with the discussion of various TCO materials, FTO films have been widely used in solar cells to replace ITO. Alternatively, FTO coated ITO/glass substrate have been proposed to overcome the shortcomings of pure ITO. FTO is the one typically used but cost-effective SnO_2-coated glass substrates on large areas (~1 m²) are still not being used as a standard substrate. On the other hand, AZO has emerged as a promising TCO material for solar cells. The AZO/glass combination has better transparency and higher conductivity than those of commercial FTO/glass substrates. Another benefit is that AZO is more resistant to hydrogen-rich plasmas used for chemical vapor deposition of thin film silicon layers as compared to FTO and ITO. The AZO films on glass for thin film silicon solar cells have a sheet resistance of about 3Ω/sq for a film thickness of ~1000 nm, a figure which degrades for thinner films. They also reported a transmittance of ~90% in the visible region of the optical spectrum for a film thickness of ~700 nm, which enhances for thinner films. These thin film silicon solar cells all have high external quantum efficiencies in the blue and green wavelength regions due to the good transmittance of the AZO films and good index matching as well as a rough interface for avoiding reflections. The highest external quantum efficiency is about 85% at a wavelength of 500 nm. However, as mentioned earlier, AZO degrades much faster than ITO and FTO in dampheat environment.

7.2 CdTe thin film solar cells

CdTe has a direct optical band gap of about 1.5 eV and high absorption coefficient of $>10^5$ cm^{-1} in the visible region of the optical spectrum, which ensures the absorption of over 99% of the incident photons with energies greater than the band gap by a CdTe layer of few micrometers in thickness. CdTe solar cells are usually fabricated in the superstrate configuration, i.e., starting at the front of the cell and proceeding to the back, as described above for the Si solar cells. CdTe is of naturally p-type conductivity due to Cd vacancies. Separation of the photo-generated carriers is performed via a CdTe/CdS p-n heterojunction. CdS is an n-type material because of native defects, and has a band gap Eg~2.4 eV, which causes light absorption in the blue wavelength range which is undesirable. For this reason, the CdS layer is made very thin and is commonly referred to as a "window layer", emphasizing that photons should pass through it to be absorbed in the CdTe "absorber layer". The basic traditional module of CdTe solar cell is composed of a stack of 'Metal/CdTe/CdS/TCO/glass'. The fabrication begins with the deposition of a TCO layer onto the planar soda lime glass sheet followed by the deposition of the CdS window layer and the CdTe light absorber layer, ~ 5 μm in thickness. Efficiencies of up to 16.5% have been achieved with small-area laboratory cells, while the best commercial modules are presently 10%–11% efficient. The thin CdS window layer poses a problem shared by both CdTe and CIS-based thin film modules, which will be discussed in the next section. Since this layer should be very thin (50–80 nm in thickness), pinholes in CdS provide a direct contact between TCO and the CdS absorber layer, creating short circuits and reducing dramatically the efficiency. This problem is especially severe for CdTe cells, because sulfur readily diffuses into the CdTe layer during post-growth annealing further decreasing the CdS layer thickness.

To mitigate this issue, thin buffer layers made of highly resistive transparent oxides are incorporated between the TCO contact and the CdS window. SnO_2 layers are commonly used as such buffers, although $ZnSnO_x$ films also have been proposed. The exact role of the buffer layers is not fully understood, whether it simply prevents short circuits by introducing resistance or also changes the interfacial energetics by introducing additional barriers, and optimization of this interface is a critical need. TCO materials typically used in CdTe solar cells are ITO and FTO. Reports for AZO in CdTe cells are very few. The use of ZnO-based TCOs in CdTe solar sells of superstrate configuration is hampered by its thermal instability and chemical reaction with CdS at high temperatures (550–650°C) typically used for CdTe solar cells fabrication. To resolve this problem, Gupta and Compaan applied low temperature (250°C) deposition by magnetron sputtering to fabricate superstrate configuration CdS/CdTe solar sells with AZO front contacts. These cells yielded efficiency as high as 14.0%. Bifacial CdTe solar cells make it possible to increase the device NIR transmission as the parasitic absorption and reflection losses are minimized. The highest efficiency of 14% was achieved from a CdTe cell with an FTO contact layer. The device performance depends strongly on the interaction between the TCO and CdS films. Later, the same group has noted a substantial In diffusion from ITO to the CdS/CdTe photodiode, which can be prevented by the use of undoped SnO_2 or ZnO buffers. Application of TCO as the back contact also allows fabrication of bifacial CdTe cells or tandem cells, which opens a variety of new applications of CdTe solar cells.

7.3 CIGS thin film solar cells

Copper indium diselenide ($CuInSe_2$ or CIS) is a direct-bandgap semiconductor with a chalcopyrite structure and belongs to a group of miscible ternary I–III–VI_2 compounds with direct optical bandgaps ranging from 1 to 3.5 eV. The miscibility of ternary compounds, that is the ability to mix in all proportions, enables quaternary alloys to be deposited with any bandgap in this range. A large light absorption coefficient of $>10^5$ cm^{-1} at photon energies greater than a bandgap allows a relatively thin (few µm in thickness) layer to be used as the light absorber. The alloy systems with optical bandgaps appropriate for solar cells include $Cu(InGa)Se_2$, $CuIn(SeS)_2$, $Cu(InAl)Se_2$, and $Cu(InGa)S_2$. Copper indium–gallium diselenide $Cu(InGa)Se_2$ (or CIGS) has been found to be the most successful absorber layer among chalcopyrite compounds investigated to date. The bandgap is ~1.0 eV for $CuInSe_2$ and increases towards the optimum value for photovoltaic solar energy conversion when gallium is added to produce $Cu(In, Ga)Se_2$. An energy bandgap of 1.25–1.3 eV corresponds to the maximum gap achievable without loss of efficiency. Further increase in the Ga fraction reduces the formation energies of point defects, primary, copper vacancies which makes them more likely to form. Also, a further increase in gallium content makes the absorber layers too highly resistive to be used in solar cells. Therefore, most CIGS devices are produced with an energy bandgap below 1.3 eV, which limits their V_{OC} at ~700 meV. Note that both CIS- and CIGS-based devices are usually dubbed as the CIS technology in the literature. The CIS technology provides the highest performance in the laboratory among all thin-film solar cells, with confirmed power conversion efficiencies of up to 20.1% for small (0.5 cm^2) cells fabricated by the Zentrum fuer Sonnenenrgie-und-Wasserstoff–Forschung and measured at the Fraunhofer Institute for Solar Energy Systems, and many companies around the world are developing a variety of manufacturing approaches aimed at low-cost, high-yield, large-area devices which would maintain laboratory-level efficiencies.

Similarly, TCO layers are generally used for the front contact, whereas a reflective contact material (Ag, frequently in combination with a TCO interlayer, is the most popular one) is needed on the back surface to enhance the light trapping in absorber layers. The optical quality of these materials substantially affects the required thickness of the absorber layers in terms of providing the absorption of an optimal amount of irradiation. Depending on the application, devices are fabricated in either a "substrate" or a "superstrate" configuration. The superstrate configuration is based on TCO-coated transparent glass substrates, and the layers are deposited in a reversed sequence, from the top (front) to the bottom (back). The deposition starts with a contact window layer of a photodiode and ends with a back reflector. Light enters the cell through the glass substrate.

In the superstrate configuration, it is important for the TCO as substrate material to be not only electrically conductive and optically transparent, but also be chemically stable during solar-cell material deposition. The superstrate design is particularly suited for building integrated solar cells in which a glass substrate can be used as an architectural element. In the case of the substrate configuration, solar cells are fabricated from the back to the front, and the deposition starts from the back reflector and is finished with a TCO layer. For some specific applications, the use of lightweight, unbreakable substrates, such as stainless steel, polyimide or PET (polyethylene terephtalate) is advantageous.

8. A novel violet and blue enhanced SINP silicon photovoltaic device

8.1 Introduction

Violet and blue enhanced semiconductor photovoltaic devices are required for various applications such as optoelectronic devices for communication, solar cell, aerospace, spectroscopic, and radiometric measurements. Silicon photodetector are sensitive from infrared to visible light but have poor responsivity in the short wavelength region. Since the absorption coefficient of crystal Si is very high for shorter wavelengths in the violet region and is small for longer wavelengths. The heavily doped emitter may contain a dead layer near the surface resulting in poor quantum efficiency of the photoelectric device under short wavelength region.

In order to improve the responsivity of silicon photodiode at the 400-600nm, a novel ITO/SiO$_2$/np Si SINP violet and blue enhanced photovoltaic device (SINP is the abbreviation of semiconductor/insulator/np structure) was successfully fabricated using thermal diffusion of phosphorus for shallow junction, a very thin silicon dioxide and ITO film as an antireflection/passivation layer. The schematic and bandgap structure of the novel SINP photovoltaic device are whown here (Fig.1 and Fig.2). The very thin SiO$_2$ film

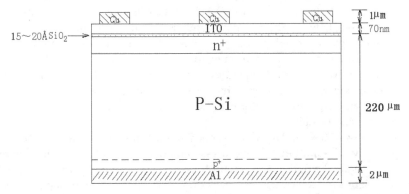

Fig. 1. Schematic of the novel SINP photovoltaic device.

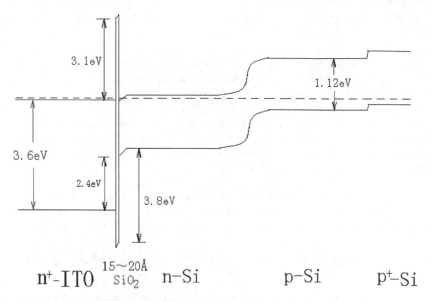

Fig. 2. Bandgap structure of the novel SINP photovoltaic device.

not only effectively passivated the surface of Si, but also reduced the mismatch of ITO and Si. Since a low surface recombination is imperative for good quantum efficiency of the device at short wavelength. The ITO film is high conducting, good antireflective (especially for violet and blue light) and stable. In addition, a wide gap semiconductor as the top film can serve as a low-resistance window, as well as the collector layer of the junction. Therefore, it can eliminate the disadvantage of high sheet resistance, which results from shallow junction. Because the penetration depth of short wavelength light is thin, the shallow junction is in favor of improving sensitivity.

8.2 Experimental in detail

The starting material was 2.0 Ωcm p-type CZ silicon. In the present, two types of shallow and deep junction n-emitters for violet and near-infrared SINP photovoltaic devices were made in an open quartz tube using liquid $POCl_3$ as the doping source. The sheet resistance is 37Ω/ $\square$ and 10Ω/ $\square$, while the junction depth is 0.35μm and 1μm, respectively. After phosphorus-silicon glass removing, a 2 μm Al metal electrode was deposited on the p-silicon as the bottom electrode by vacuum evaporation. The 15~20Å thin silicon oxide film was successfully grown by low temperature thermally (500°C for 20 min in N_2:O_2=4:1 condition) grown oxidation technology. The 70 nm ITO antireflection film was deposited on the substrate in a RF magnetron sputtering system. Sputtering was carried out at a working gas (pure Ar) pressure of 1.0Pa.

The Ar flow ratio was 30 sccm. The RF power and the substrate temperature were 100W and 300°C, respectively. The sputtering was processed for 0.5h.The ITO films were also prepared on glass to investigate the optical and electrical properties. Finally, by sputtering, a 1μm Cu metal film was deposited with a shadow mask on the ITO surface for the top grids electrode. The area of the device is 4.0 cm^2.

8.3 Results and discussion
8.3.1 Optical and electric properties of ITO films
In order to learn the optical absorption and energy band structure of ITO film, the transmission spectrum of the ITO film deposited on the glass substrate was measured (Fig.3). The thickness of ITO film is about 700 Å. The average transmittance of the film is about 95% in the visible region and the band-edge at 325nm.While the optical band gap of ITO film is about 3.8 eV by calculation. The reflection loss for ITO film on a texturized Si surface was indicated (Fig.4) from UV to the visible regime, which is much lower than that of Si$_3$N$_4$ film that are widely made by PECVD technology. This shows that ITO film effectively reduced reflection loss in short-wavelength, which is suitable for antireflection

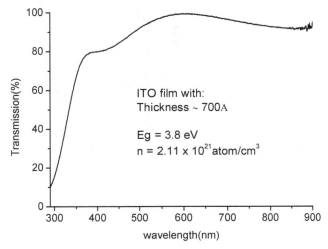

Fig. 3. Transmission spectrum of the ITO film.

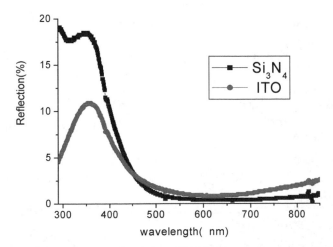

Fig. 4. Comparison of the reflections for ITO and Si$_3$N$_4$ films on a texturized Si surface.

coating in violet and blue photovoltaic device. Electrical properties of the ITO film were measured by four-point probe and Hall effect measurement. The square resistance and the resistivity are low to 17Ω/□ and 1.19×10^{-4} Ω·cm, respectively, while carrier concentration is high to 2.11×10^{21} atom/cm³.

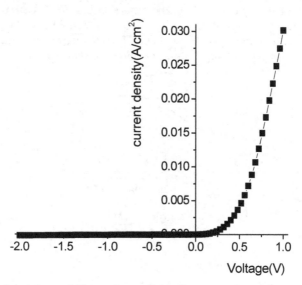

Fig. 5. I-V curve of the violet and blue enhanced (shallow junction) SINP photovoltaic device in dark.

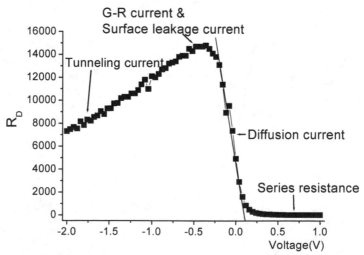

Fig. 6. The variation of resistance for SINP violet device via voltage (R_D-V curve).

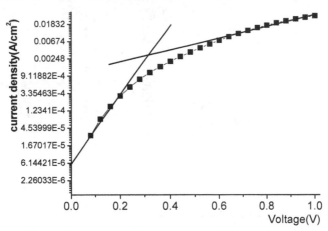

Fig. 7. The corresponding logarithmic scale in current with forward bias condition.

8.3.2 I-V characteristics

In our study, the current-voltage characteristic of the violet SINP device was measured in dark at room temperature (in Fig.5). I-V curves of the devices show fairly good rectifying behaviors. Basing on the dark current as a function of the applied bias, the corresponding diode resistance defined as $R_D = (\frac{dI}{dV})^{-1}$ is derived and shown (in Fig.6). The series resistance arose from ohmic depletion plays a dominant role when the forward bias is larger than 0.25 V. When the voltage varies within 0.2 V and - 0.2 V, the resistance slightly increases as the diffusion current in the base region. When the inversion voltage increases from - 0.2 to - 0.5 V, the leakage current and the recombination current in the surface layers restrain the increase of the dynamic resistance, which keeps the R_D – V curve in an invariation state. In the high inversion voltage region, the tunneling current plays a dominant role.

The plot of $\ln(J)$ against V, is shown (in Fig.7), which indicates that the current at low voltage (V < 0.3 V) varies exponentially with voltage. The characteristics can be described by the standard diode equation: $J = J_0(e^{\frac{qV}{nk_BT}} - 1)$ where q is the electronic charge, V is the applied voltage, k_B is the Boltzmann constant, n is the ideality factor and J_0 is the saturation current density. Calculation of J_0 and n from is obtained the measurements (in Fig.7). The value of the ideality factor of the violet SINP device is determined from the slop of the straight line region of the forward bias log(I)-V characteristics. At low forward bias (V< 0.2 V), the typical values of the ideality factors and the reverse saturation current density are 1.84 and 5.58×10^{-6} A/cm^2, respectively.

Using the standard diode equation $J = J_0(e^{\frac{qV}{nk_BT}} - 1)$, where n = 1.84 and J_0 = 5.58×10^{-6} A/cm^2. The result of calculation is similar to that of the measurement (in I-V curve). By the same calculation method, the ideality factor and the reverse saturation current density of deep junction SINP photovoltaic device are 2.21 and 4.2×10^{-6} A/cm^2, respectively. This result indicates that the recombination current $J_r \approx \exp(qV/2kT)$ dominates in the forward current. The rectifying behaviors and the composition of dark current for violet SINP photovoltaic device is better than deep junction SINP device, because the ideality factor of the violet SINP

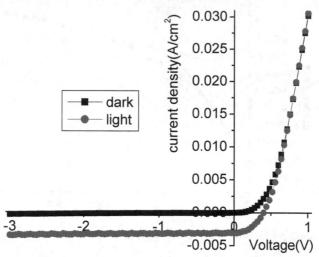

Fig. 8. I-V characteristic of the violet and blue enhanced SINP photovoltaic devices in dark and light (6.3 mW/cm² - white light), respectively.

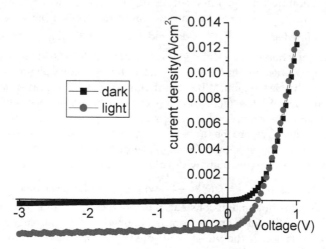

Fig. 9. I-V characteristic of the deep junction SINP devices in dark and light (6.3 mW/cm² - white light), respectively.

photovoltaic device (n=1.84) is lower than that of the deep junction SINP device (n=2.21). Furthermore, the values of I_F/I_R (I_F and I_R stand for forward and reverse current, respectively) at 1V for violet SINP device and deep junction SINP device are found to be as high as 324.7 and 98.4, respectively.

The weak light-injection I-V characteristics of the novel SINP devices with low power white light (6.3mW/cm²) illuminating were measured at 23°C. It is observed that the novel SINP device exhibits a good photovoltaic effect and rectifying behavior in the photon – induced carrieres transportation. On the other side, another essential physical parameter is internal quantum efficiency (IQE) or external quantum efficiency (EQE) for the evaluation of the spectra response of the light (Fig.8 and Fig.9). The photocurrent density ($\sim 3.08 \times 10^{-3}$ A/cm²) of violet and blue enhanced SINP photovoltaic device is much higher than that of deep junction SINP device ($\sim 2.23 \times 10^{-3}$A/cm²), at V = 0.

8.3.3 Spectral response and responsivity

The comparison of IQE, EQE and the responsivity for the violet and blue SINP photovoltaic device and the deep junction SINP photovoltaic device has been illustrated (in Fig.10 ~ Fig.12). In visible light region, the internal and external quantum efficiencies (IQE and EQE) of the devices are in the range of 75% to 85%. In the violet and blue region, the IQE and EQE of shallow junction violet SINP device is much higher than that of the deep junction SINP device. For example, the EQE and the responsivity of the violet SINP device are 70% and 285mA/W at 500nm, respectively, while the EQE and the responsivity of the deep junction SINP device are 42% and 167mA/W at 500nm, respectively. The spectral responsivity peak of violet and blue SINP photovoltaic device is 487mA/W at about 800nm. While the spectral responsivity peak of deep junction SINP photovoltaic device is 471mA/W at about 860nm. The high quantum efficiency and the responsivity of violet and blue enhanced photovoltaic cell attribute to the shallow junction and the good conductive, and the violet and blue antireflection of ITO film.

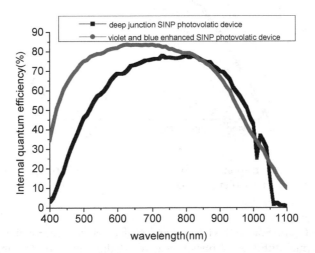

Fig. 10. Comparison of IQE for violet and blue SINP photovoltaic device and the deep junction SINP photovoltaic device.

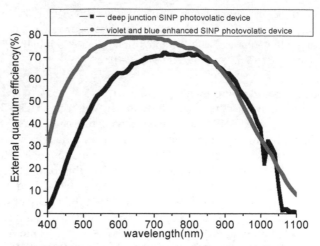

Fig. 11. Comparison of EQE for violet and the blue SINP photovoltaic device and the deep junction SINP photovoltaic device.

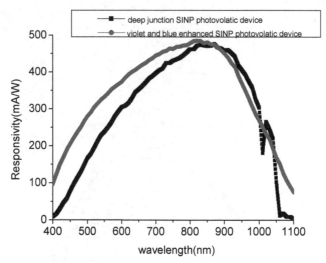

Fig. 12. Comparison of the responsivity for the violet and blue SINP photovoltaic device and the deep junction SINP photovoltaic device.

8.3.4 Conclusions

The novel ITO/SiO$_2$/np Silicon SINP violet and blue enhanced photovoltaic device has been fabricated by thermal diffusion of phosphorus for shallow junction to enhance the spectral responsivity within the wavelength range of 400-600nm, the low temperature thermally grown a very thin silicon dioxide and RF sputtering ITO antireflection coating to reduce the reflected light and enhance the sensitivity. The ITO film was evinced to a high quality by UV-VIS spectrophotometer, four point probe and Hall-effect measurement. Fairly good rectifying and obvious photovoltaic behaviors are obtained and analyzed by I-V measurements. The spectral response and the responsivity with a higher quantum efficiency of the violet SINP photovoltaic device and the deep junction SINP photovoltaic device were analyzed in detail. The results indicated that the novel violet and blue enhanced photovoltaic device could be not only used for high quantum efficiency of violet and blue enhanced silicon photodetector for various applications, but also could be used for the high efficiency solar cell.

9. Fabrication and photoelectric properties of AZO/SiO₂/p-Si heterojunction device

9.1 Introduction

As shown in the previous work, semiconductor-insulator-semiconductor (SIS) diodes have certain features, which make them more attractive for the solar energy conversion than conventional Shottky, MIS, or other heterojunction structures (Mridha et al., 2007). For example, efficient SIS solar cells such as indium tin oxide (ITO) on silicon have been reported, where the crystal structures and the lattice parameters of Si (diamond, a = 0.5431 nm), SnO_2 (tetragonal, a = 0.4737 nm, c = 0.3185 nm), In_2O_3 (cubic, a = 1.0118 nm) show that they are not particularly compatible and thus not likely to form good devices. However, the SIS structure is potentially more stable and theoretically more efficient than either a Schottky or a MIS structure. The origins of this potential superiority are the suppression of majority-carrier tunneling in the high potential barrier region of SIS structure, and the existence of thin interface layer which minimizes the amount and the impact of the interface states. This results in an extensive choice of the p-n junction partner with a matching band gap in the front layer. In addition, the top semiconductor film can serve as an antireflection coating (Dengyuan et al., 2002), a low-resistance window, and the collector of the p-n junction as well.

Furthermore, the semiconductor with a wide band gap as the top layer of SIS structure can eliminate the surface dead layer which often occurs within the homojunction devices, such as the normal bulk silicon based solar cells. On the other side, this absence of the light absorption of visible region in a surface layer can improve the ultraviolet response of the internal quantum efficiency. Among many transparent conductive oxides (TCO) of the transition metals, ZnO:Al is one the best n-type semiconductor layer. It has high conductivity, high transmittance, optimized surface texture for light trapping, and large band gap of $E_g \approx$ 3.3 eV. Thus, in this description, we show a photovoltaic device with AZO/SiO₂/p-Si frame, as an attempt to study its opto-electronic conversion property and the I-V features as well. The schematic and the bandgap structure of the novel AZO/SiO₂/p-Si SIS heterojunction device was show here (Fig.13).

9.2 Experimental in details

For the purpose of fabricating SIS structure, p-type Si (100) wafers were used as the substrates of the heterojunction device. The wafers were firstly prepared by a stand cleaning procedure, then, they were dipped in 10% HF solution for one minute to remove native oxide layer. Finally, the wafers were dried in a flow gas of nitrogen.

By thermal evaporation, 1 μm-thick Al electrode was deposited on the back side. Then the samples were annealed at 500°C for 20 min in $N_2:O_2$=4:1 condition to form good ohmic contact and a very thin oxide layer (about 15~20Å) was grown on the p-Si surface.

The Al doped ZnO films were deposited on the oxidized silicon substrates in a RF magnetron sputtering system. The target was a sintered ceramic disk of ZnO doped with 2 wt% Al_2O_3 (purity 99.99%). The base pressure inside the chamber was pumped down to less than 5×10^{-4} Pa. Sputtering was carried out at a working gas (pure Ar) pressure of 1Pa. The Ar flow ratio was 30 sccm. The RF power and the temperature on substrates were kept at 100W and 300°C, respectively. The sputtering was proceeded for 2.5 hours. The area is 2×2 cm².

The thickness of AZO film was measured by step profiler. The optical transmission of the films was measured by UV-VIS spectrophotometer. The electrical properties of Al doped ZnO films were characterized by four point probe. The current-voltage characteristics of the device was measured by Agilent 4155C semiconductor parameter analyzer (with probe station, the point diameter of a probe is 5 μm).

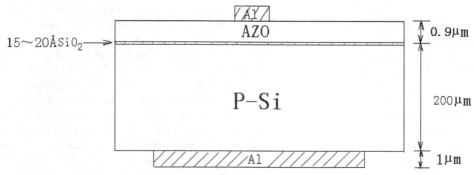

Fig. 13. The structure of $AZO/SiO_2/$p-Si heterojunction PV device.

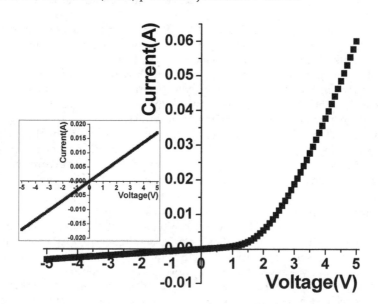

Fig. 14. I-V curve of the $Al/AZO/SiO_2/$p-Si/Al heterojunction device in dark.

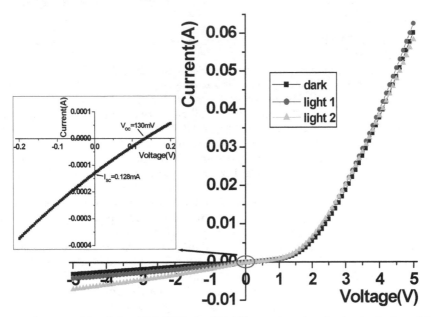

Fig. 15. I-V characteristic of the AZO/SiO₂/p-Si/Al heterojunction device in dark and light (light-1: 6.3mW/cm² white light; Light-2: 20W halogen lamp)

9.3 I-V characteristics

A linear I-V behavior between the two electrodes on the surface of ZnO:Al film indicates a good ohmic contact. The current-voltage characteristic of the AZO/SiO₂/p-Si/Al heterojunction device was measured at room temperature in the dark (Fig.14). Typical rectifying is observed for this heterojunction with polar to covalent semiconductors structure. The weak photon irradiation I-V characteristics were measured under two kinds of illumination by low power white light (6.3mW/cm²) lamp and 20W halogen lamp (in Fig.15). The good rectifying with the increase of photoelectric current was observed for the typical interface mismatching device. Under reverse bias conditions the photocurrent caused by the ZnO surfaces exposing in the low power white light lamp and 20W halogen lamp was obviously much larger than the dark current. For example, when the reverse bias is -5V, the dark current is only 3.05×10^{-3}A.While the photocurrent reach to 4.06×10^{-3}A and 6.99×10^{-3}A under low power white light and halogen lamp illumination, respectively.

9.4 Conclusions

The novel AZO/SiO₂/p-Si/ heterojunction has been fabricated by magnetron sputtering deposition AZO film on p-Si substrate. Fairly good rectifying and photoelectric behaviors are observed and analyzed by I-V measurements in detail. The ideality factor and the saturation current of this diode is 20.1 and 1.19×10^{-4}A, respectively. The results indicated that the novel AZO/SiO₂/p-Si/ heterojunction device could be not only used for low cost solar cell, but also could be used for the high quantum efficiency enhanced photodiode in UV and visible lights, and also for other applications.

10. References

Ashok, S.; Sharma, P.P.; Fonash, S.J. (1980) IEEE Transactions on Electron Devices 27, (1980) pp. 725-730

Bruk, L.; Fedorov, V.; Sherban, D.; Simashkevich, A.; Usatii, I.; Bobeico, E.; Morvillo, P. (2009) "Isotype bifacial silicon solar cells obtained by ITO spray pyrolysis", Materials Science and Engineering B 159-160 (2009) pp.282-285

Baik, D.G.; Cho, S.M. (1999). "Application of sol-gel derived films for ZnO/n-Si junction solar cells, Thin Solid Films 354 (1999) pp.227-231

Chaabouni, F.; Abaab, M. Rezig, B. (2006). "Characterization of n-ZnO/p-Si films grown by magnetron sputtering", Superlattices and Microstructures, vol 39, (2006) pp.171-178

Cheknane, Ali (2009) "Analytical modelling and experimental studies of SIS tunnel solar cells", J. Phys. D: Appl. Phys. 42 (2009) pp.115302-115307

Chen, X. D.; Ling, C. C.; Fung, S.; Beling, C. D. (2006). "Current transport studies of ZnO/p-Si heterostructures grown by plasma immersion ion implantation and deposition", Applied Physics Letters, vol 88, (2006) pp.132104-132107

Canfield, L.R.; Kerner, J.; Korde, R. (1990). "Silicon photodiodes optimized for the EUV and soft xray regions", EUV.X-Ray.andGamma-Ray Instrumentation for Astronomy. SPIE vol 1344 , (1990) pp. 372−377

Granqvist, C.G. (2007) "Transparent conductors as solar energy materials: A panoramic review", Sol. Ener. Mater. Sol. Cells 91, (2007) pp.1529-1598

Granqvist, Claes G. (2007) "Transparent conductors as solar energy materials: A panoramic review", Solar Energy Materials & Solar Cells 91 (2007) pp.1529–1598

He, B.; Ma, Z.Q. et al., (2009). "Realization and Characterization of ITO/AZO/SiO$_2$/p-Si SIS Heterojunction", Superlattices and Microstructures 46, (2009) pp.664-671

He, B.; Ma, Z.Q. et al., (2010). "Investigation of ultraviolet response enhanced PV cell with silicon-based SINP configuration", SCIENCE CHINA Technological Sciences 53, (2010) pp.1028-1037

Koida, T.; Fujiwara, H.; Kondo, M. (2009) "High-mobility hydrogen-doped In$_2$O$_3$ transparent conductive oxide for a-Si:H/c-Si heterojunction solar cells", Solar Energy Materials & Solar Cells 93 (2009) pp.851–854

Kittidachachan, P.; Markvart, T.; Ensell, G.J.; Greef, R.; Bagnall, D.M. (2005). "An analysis of a "dead layer" in the emitter of n+pp+ solar cells", Photovoltaic Specialists Conference, Conference Record of the Thirty-first IEEE, vol 3-7, (Jan., 2005) pp. 1103-1106

Mridha, S.; Basak, Durga. (2007). "Ultraviolet and visible photoresponse properties of n-ZnO/p-Si heterojunction", Journal of Applied Physics, Vol 101, (2007) pp.083102

Ramamoorthy, K.; Kumar, K.; Chandramohan, R.; Sankaranarayanan, K. (2006) "Review on material properties of IZO thin films useful as epi-n-TCOs in opto-electronic (SIS solar cells, polymeric LEDs) devices", Materials Science and Engineering B 126 (2006) pp.1–15

Song, Dengyuan; Aberle, Armin G.; Xia, James (2002). "Optimisation of ZnO:Al films by change of sputter gas pressure for solar cell application", Applied Surface Science 195, (2002) pp.291-296

Singh, R.; Rajkanan, K.; Brodie, D.E.; Morgan, J.H. (1980) IEEE Transactions on Electron Devices 27, (1980) pp. 656-662

Wenas, Wilson W.; Riyadi, S. (2006). "Carrier transport in high-efficiency ZnO/SiO$_2$/Si solar cells", Solar Energy Materials and Solar Cells 90 , (2006) pp.3261-3267

8

Architectural Design Criteria for Spacecraft Solar Arrays

Antonio De Luca
VEGA Space GmbH
Germany

1. Introduction

Scope of this chapter is to provide design criteria for spacecraft solar arrays at system level.
The design a satellite solar array is usually influenced by several constraints; mission profile, chosen attitude, overall spacecraft configuration, mass and sizing requirements, etc.
Moreover, its design has to be harmonised with the chosen solar array power conditioning, in order to optimize mass, dimensions, and also particular constraints coming from EMC and thermal environments.
The chapter is basically composed of the following sections;

1. General description of the current solar cell technologies currently used in space, with particular attention to the triple junction solar cells.
2. Mathematical model of an equivalent solar cell circuit, to be used for performance calculations in a numerical simulation environment.
3. Mathematical description of a simplified thermal model of a solar array in order to analyse solar array performances in orbit.
4. Short definition of cosmic radiation effects.
5. The satellite power budget, starting point for the solar array sizing
6. The impact of the power conditioning architecture on the solar array (electrical operative point, EMC considerations).
7. The configuration of the solar array with respect to the spacecraft.
8. Some design examples for different missions and satellite configurations.
9. Numerical simulations of solar array performances as function of the mission profile (orbit propagation, slew manoeuvres, attitudes of particular interest).

2. Solar cells for space applications

Since the beginning of the astronautic era, photovoltaic devices have been considered for the generation of electrical power on board spacecrafts because of their high power output per unit mass, associated with the fundamental advantage of not having moving parts, present, instead, in all the most used electrical power generators for both terrestrial and aeronautical applications (turbines, motors, alternators, etc.). Therefore the PV array is static, does not produce vibrations or noise, and does not need an active cooling. The Russians were the first, in 1958, to launch a satellite powered with silicon solar cells.

Solar cells for space applications have to be highly efficient, capable to stand thousands of thermal cycles in orbit where the temperature, according to the mission profile may vary from -150 °C to more than 120 °C. They have to show a limited degradation during time due to cosmic radiations and Ultraviolet, and they have to resist to the mechanical solicitations mainly linear accelerations and vibrations during launch and orbital manoeuvres, because of these constraints the cells for space are smaller than those for terrestrial applications.

In order to have the highest conversion efficiency, solar cells for space application are developed from mono-crystalline materials. In the past silicon was the most used and the reachable bulk efficiency was not higher than 14%. The advent of GaAs based solar cells in the last decade of the 20th century took the efficiency up to 19%, and nowadays triple junction solar cells show more than 30%.

Figure 1 shows a very simplified structure of triple junction cell.

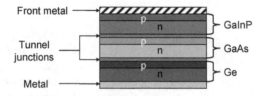

Fig. 1. Triple junction solar cell structure

While figure 2 reports the quantum efficiency for each junction, it can be clearly seen that the increased efficiency is due to wider wavelength coverage of the absorbed radiation.

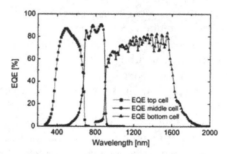

Fig. 2. Equivalent quantum efficiency as function of wavelength

Triple junction GaAs solar cells are populating more and more solar generators worldwide, while manufacturers are actively working on four to six junction cells as a way forward always increasing conversion efficiency. Consequently, there is a need to improve the understanding of the electrical dynamic behaviour of multi-junction based solar array considering that the proper design of solar array regulators requires, among others, a good mastering of the solar section/regulator interface. In order to better understand EMC aspects connected to the chosen regulation philosophy, which will be discussed further, it is worth to have a quick look at the equivalent capacitance present at the output of a triple junction cell. The figure 3 reports the capacitance measured across strings composed of 15 cells. The cells used are produced by AZUR SPACE Solar Power GmbH. It can be observed that at high voltages the capacitance is considerably increased. Such behaviour has to be taken into account when the power conditioning architecture is chosen, and the relevant devices designed.

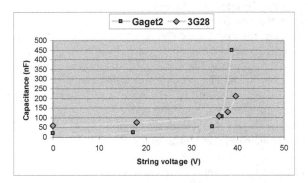

Fig. 3. Capacitance identified for the 15 cells string, Gaget2 and 3G28 (AZUR SPACE products)

3. Solar cell equivalent circuit

The mathematical model of a photovoltaic cell has to take into account the following factors capable to influence the solar cell behaviour.
1. Intensity of the incident light.
2. Operative absolute temperature.
3. Degradation by cosmic radiation.

The solar cell model, derived from the Mottet-Sombrin's one, is basically a current generator driven by the value of the voltage applied at its terminal according to the equivalent circuit reported below. Generally speaking a solar cell is a particular p-n junction where the diffusion process (diode D1) co-exists with the generation and recombination effect of the charge carrier (diode D2) induced by the presence of crystalline defects. This model was tested using data relevant to the AZUR SPACE 28% solar cell, as reported in the datasheet provided by the Manufacturer, and available on company web-site.

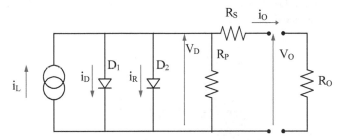

Fig. 4. Equivalent Circuit of solar cell

The relevant Kirchhoff equations are:

$$i_o = i_L - i_D \cdot \left[\exp\left(\frac{q \cdot V_D}{k \cdot T}\right) - 1 \right] - i_R \cdot \left[\exp\left(\frac{q \cdot V_D}{2k \cdot T}\right) - 1 \right] - \frac{V_D}{R_p} \qquad (1a)$$

$$V_o = V_D - R_S \cdot i_o \tag{1b}$$

Where:

K=1.381×10⁻²³ (J/°K) is the Boltzmann constant;

q=1.602×10⁻¹⁹ (C) is the electron charge;

i_L, i_D e i_R are respectively the current due to illumination, and the reverse currents of the diodes D_1 e D_2; they are function of the temperature.

The equations (1) give the output voltage V_o, and current I_o as function of the voltage drop V_d over the diodes D_1 and D_2. The second and third term of (1_a) represent the typical voltage-current laws of the diodes, and the currents i_D and i_R are the reverse currents of the diodes dependent from the physics of the solar cell.

In general, the solar cell is characterised by the following data provided in the manufacturer's data sheet, the table below gives the values relevant to the one used for testing the model:

I_{sc}	506.0 mA	Short circuit current;
I_{mp}	487.0 mA	Maximum power current;
V_{mp}	2371.0 mV	Maximum power voltage;
V_{oc}	2667.0 mV	Open circuit voltage;
dI_{sc}/dT	0.32 mA/°K	Short circuit current temperature coefficient
dI_{mp}/dT	0.28 mA/°K	Max. power current temperature coefficient
dV_{mp}/dT	-6.1 mV/°K	Max. power voltage temperature coefficient;
dV_{oc}/dT	-6.0 mV/°K	Open circuit voltage temperature coefficient.

Such data are given in AM0 (1367.0 W/m²) conditions at T_{ref}=28 °C (301.15 °K) reference temperature.

Usually the series resistance is around 300mΩ for a triple junction cell, while for the shunt one 500Ω maybe assumed. Such resistances may be considered in a first approximation as constant in the operating temperature range of the cell.

The values of i_L, i_D and i_R at the reference temperature can be calculated with the (1) in the three main point of the V-I curve; short circuit, maximum power and open circuit, by the least square method.

The next step is to define how these currents change with temperature.

Concerning i_D e i_R it is possible to write:

$$i_D = C_D \cdot T^{5/2} \cdot \exp\left(\frac{E_g}{n_1 \cdot T}\right) \tag{2a}$$

$$i_R = C_R \cdot \exp\left(\frac{E_g}{n_2 \cdot k \cdot T}\right) \tag{2b}$$

Where C_D and C_R are constants independent from temperature, and E_g is the Energy of the prohibited band gap:

$$E_g = E_{g0} - \frac{\left(\alpha_e \cdot T^2\right)}{\left(T + \beta_e\right)} \quad (mA/cm^2) \tag{3}$$

With E_{g0} = 1.41 eV, α_e=-6.6×10⁻⁴ eV/°K, and β_e=552 °K.
The current i_L due to illumination is given instead by

$$i_L(T) = K(T) \cdot \eta(T) \cdot J_{tot} \quad (mA/cm^2) \tag{4}$$

Where J_{tot} is light intensity (W/ m²), $\eta(T)$ is the efficiency of the cell, $K(T)$ is a coefficient to be determined as function of the temperature.
n_1 e n_2 are two coefficients depending on the adopted solar cell technology:
At this point all the terms of the equations (1) can be defined at any temperature and by setting as input the operating voltage V_o and solving the system by the Newton-Raphson numerical scheme is possible to calculate the output current i_o.
Figure 5 shows the V-I curves relevant to Triple Junction AZUR SPACE solar cell starting from the datasheet available on the web site, as function of temperature at Begin Of Life (BOL); the black asterisks are the maximum power points calculated according to the datasheet. In figure 6 V-I curves for different illumination levels are reported.

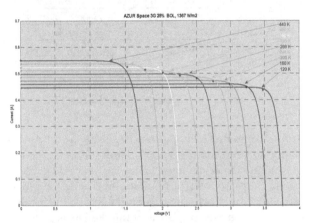

Fig. 5. Computed V-I curves as function of temperature using AZUR SPACE 3G 28% data

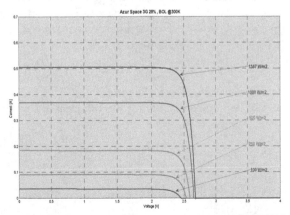

Fig. 6. Computed V-I curves as function of illumination using AZUR SPACE 3G 28% data

4. Solar panel thermal model

What said above clearly highlights the need of a thermal model of the solar panel taking into consideration the heat exchange on both sides of it in case of a deployable one; or usually considering the rear side as adiabatic in case of the panel is body mounted. The panel is considered as rigid, with honeycomb structure on which the solar cells are applied; the following table reports the components recognizable in solar panel cross-section:

Components from front to rear side	Thickness, μm
Coverglass	150 - 500
Coverglass adhesive	50 - 100
Solar cell	100 - 200
Cell adhesive	100
Kapton insulation	50
Face skin (Carbon fibre)	-
Adhesive	100
Honeycomb (Aluminium)	Up to 4 cm
Adhesive	100
Face skin (Carbon fibre)	-
Black paint	50

Table 1. Solar panel composition

The panel temperature is computed taking into account the direct sun radiation, the albedo radiation, the irradiation to deep space, and irradiation between the earth surface and the panel itself. The sun illumination is variable during the year and considering only missions around the earth it may range between 1315.0 (summer solstice) and 1426 W/m2 (winter solstice), while the albedo of the earth surface is about 30% of the incident sun illumination. The panel exchanges heat with the deep space and this is seen as a black body at 3°K, as well as the earth irradiates as a black body at 250°K. The following simplifying assumptions can been made; a deployed solar panel does not exchange heat with the outer surfaces of the satellite body, a body mounted solar panel is adiabatically isolated from the rest of the satellite body and finally the panel surface temperature is considered as uniform. The conduction across the panel also plays an important role, and it has to be taken into account in case of a deployed solar panel. At a first glance, the in-plane conductivity may be neglected, this because under the hypothesis of uniform temperature over the panel, the heat exchange between adjacent cells is basically zero.

The thermal equilibrium is computed by solving the differential equation which takes into account the different heat exchange modalities.

$$C \cdot \frac{dT}{dt} = Q_{Rad} + Q_{Alb} + Q_{Earth} + Q_{Space} + Q_{cond} \tag{5}$$

Where:

$$Q_{Rad} = \alpha (1 - \eta) \cos \vartheta \cdot J \tag{6}$$

It is the contribution of the direct sun radiation J that is not converted into electrical power by the photovoltaic cell;

$$Q_{alb} = \alpha \left(1 - \eta\right) \cdot F \cdot Al \cdot F_{1-2} \cdot J \tag{7}$$

It is the contribution of the albedo radiation;

$$Q_{Earth} = \sigma \cdot \varepsilon \cdot F_{12} \cdot \left(T_E^4 - T^4\right) \tag{8}$$

It is the heat exchanged with the Earth surface.

$$Q_{Space} = \sigma \cdot \varepsilon \cdot \left(1 - F_{12}\right) \cdot T^4 \tag{9}$$

It is the heat released to the deep space.

$$Q_{cond} = \frac{\left(T_{front} - T_{rear}\right)}{\Delta x} \cdot k \tag{10}$$

Is the heat transmitted by conduction between the front and rear faces of the panel at T_{front} and T_{rear} temperature respectively.

The parameters appearing in these equations have the following meanings:

C = thermal capacitance of the panel per unit area, main contribution is provided the honeycomb structure;

α = solar cell absorptivity;

ε = solar cell emissivity;

Al = albedo coefficient, about 0.3 for earth;

F = albedo visibility factor;

F_{12} = View factor between radiating surface and planet

σ = Stephan-Boltzmann, constant: 5.672×10^{-8} W/(m$^2 \times$°K^4);

T_E = Black body equivalent temperature of the earth;

ϑ = incidence angle of the sunlight on the panel;

k = panel transverse thermal conductivity;

Δx = panel thickness;

The radiating view factor of a flat surface with respect to the Earth surface is function of the altitude h, earth radium $R_\oplus$ and the angle λ between the nadir and the normal to the panel. The albedo view factor is computed according to the following formulas:

$$F_{alb} = \left[1 - \cos\left(\beta_{app}\right)\right] \cdot \cos\left(\beta\right) \tag{11}$$

Where β is the angle between the nadir and the earth-sun direction:

$$\beta_{app} = \arcsin\left(\frac{R_\oplus}{h + R_\oplus}\right) \tag{12}$$

The integration in the time domain of the equation (5) gives the actual temperature of the panel along the propagation of the orbit in eclipse and sunlight, taking into account the orientation of the panel itself with respect to the earth and the sun. The thermal model exposed so far is sufficient for the design of a solar array for space application at system level.

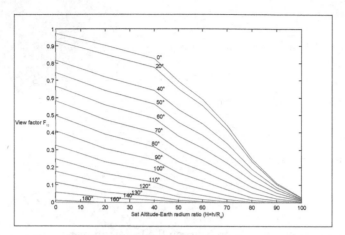

Fig. 7. View Factors F_{12} as function of H=h/$R_\oplus$, parameter:λ

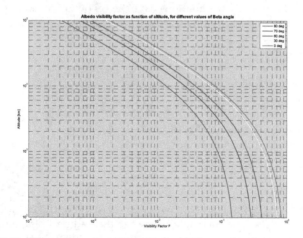

Fig. 8. Albedo Visibility factor F as function of h for different β value

5. Degradation due to space environment

Space is a hostile environment for the electronic components in general and solar cells in particular. The sun radiates energy in almost the whole electromagnetic spectrum, from radio waves to gamma rays, and abundant charged particles impinging on a surface cause damages which cumulate over the mission lifetime (fluence Φ). The behaviour of solar cells in a radiation environment can be described in terms of the changes in the engineering output parameters of the devices. The radiation usually of interest in the study of degradation of materials and devices consists of energetic or fast massive particles (i.e. electrons, protons, neutrons or ions). The major types of radiation damage phenomena in solids which are of interest to the solar array designer are ionisation and atomic displacement.

Ionisation occurs when orbital electrons are removed from an atom or molecule in gases, liquids, or solids. The measure of the intensity of ionising radiation is the roentgen. The measure of the absorbed dose in any material of interest is usually defined in terms of absorbed energy per unit mass. The accepted unit of absorbed dose is the rad (100 erg/g or 0.01 J/kg). For electrons, the absorbed dose may be computed from the incident fluence Φ (in cm^{-2}) as: Dose (rad) = 1.6×10^{-8} dE/dx Φ, where dE/dx (in MeV cm^2 g^{-1}) is the electron stopping power in the material of interest. In this manner, the effects of an exposure to fluxes of trapped electrons of various energies in space can be reduced to an absorbed dose. By the concept of absorbed dose, various radiation exposures can be reduced to absorbed dose units which reflect the degree of ionisation damage in the material of interest. This concept can be applied to electron, gamma, and X-ray radiation of all energies. Several ionisation related effects may degrade the solar cell assemblies. The reduction of transmittance in solar cell cover glasses is an important effect of ionising radiation.

The basis for solar cells damage is the displacement of semiconductor atoms from their lattice sites by fast particles in the crystalline absorber. The displaced atoms and their associated vacancies after various processes form stable defects producing changes in the equilibrium of carrier concentrations and in the minority carrier lifetime. Such displacements require a certain minimum energy similar to that of other atomic movements. Seitz and Koehler [1956] estimated the displacement energy is roughly four times the sublimation energy. Electron threshold energies up to 145 keV have been reported. Particles below this threshold energy cannot produce displacement damage, therefore the space environment energy spectra are cut off below this value. The basic solar cell equations (1) may be used to describe the changes which occur during irradiation. This method would require data regarding the changes in the light generated current, series resistance, shunt resistance, but most investigations have not reported enough data to determine the variations in the above parameters. The usual practice is then to reduce the experimental data in terms of changes in the cell short circuit current (I_{sc}), open circuit voltage (V_{oc}), and maximum power (P_{max}). The variation of common solar cell output parameters during irradiation can be described as shown for I_{sc} in the following case:

$$I_{sc} = I_{sc0} - C \log (1 + \Phi / \Phi_x) \tag{13}$$

Where Φ_x represents the radiation fluence at which I_{sc} starts to change to a linear function of the logarithm of the fluence. The constant C represents the decrease in I_{sc} per decade in radiation fluence in the logarithmic region. In a similar way, for the V_{oc} it can be written;

$$V_{oc} = V_{oc0} - C' \log (1 + \Phi / \Phi_x). \tag{14}$$

And for the maximum power;

$$P_{max} = P_{max0} - C'' \log (1 + \Phi / \Phi_x). \tag{15}$$

In the space environment a wide range of electron and proton energies is present; therefore some method for describing the effects of various types of radiation is needed in order to get a radiation environment which can be reproduced in laboratory. It is possible to determine an equivalent damage due to irradiation based upon the changes in solar cell parameters which are in some way related to the minority carrier diffusion length.

The I_{sc} variation in each environment is described by the equation for I_{sc}. In this case, two constants, C and Φ_x, are required to describe the changes in I_{sc}. It has been shown that the constant C, under solar illumination, does not greatly vary for different radiation environments. For electron irradiations in the 1 MeV and greater range, C is about 4.5 to 5.5 mA cm^{-2}/decade. In case of proton and neutron, C approaches 6 to 7 mA cm^{-2}/decade.

For solar cells with the same initial I_{sc}, the constant Φ_x is a measure of the damage effectiveness of different radiation environments. The constant Φ_x for a particular radiation can be determined graphically on a semi-log plot at the intersection of the starting I_{sc} and the extrapolation of the linear degradation region.

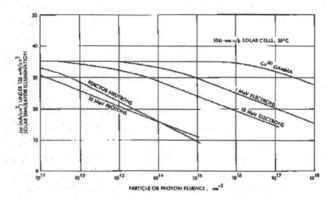

Fig. 9. Variation of solar cell short circuit current with fluence for various radiations

It is the practice to define an arbitrary constant referred to as the critical fluence Φ_c. One method of defining this value is that fluence which degrades a solar cell parameter 25% below its BOL state. But such a parameter is valid only when comparing cells with similar initial parameters. To eliminate this problem, critical fluence may be alternatively defined as that fluence which will degrade a cell parameter to a certain value. By use of the critical fluence or the diffusion length damage coefficient, it is possible to construct a model in which the various components of a combined radiation environment can be described in terms of a damage equivalent fluence of a selected mono-energetic particle. 1 MeV Electrons are a common and significant component of space radiation and can be produced conveniently in a test environment. For this reason, 1 MeV electron fluence has been used as a basis of the damage equivalent fluences which describe solar cell degradation.

The degradation due to radiation effects on solar cell cover-glass material in space is difficult to assess. The different radiation components of the environment act both individually and synergistically on the elements of the shielding material and also cause changes in the interaction of shielding elements. However, the most significant radiation effects in cover materials involve changes in the transmission of light in the visible and near infrared region.

The methods for estimating solar cell degradation in space are based on the techniques described by *Brown et al.* [1963] and *Tada* [1973ab]. In summary, the omni-directional space radiation is converted to a damage equivalent unidirectional fluence at a normalised energy and in terms of a specific radiation particle. This equivalent fluence will produce the same damage as that produced by omni directional space radiation considered if the relative damage coefficient (RDC) is properly defined to allow the conversion. When the equivalent fluence is determined for a given space environment, the parameter degradation can be evaluated in the laboratory by irradiating the solar cell with the calculated fluence level of unidirectional normally incident flux. The equivalent fluence is normally expressed in terms of 1 MeV electrons or 10 MeV protons. The three basic input elements necessary to perform degradation calculations are:

1. degradation data for solar cells under normal incidence 1 MeV electron irradiation;
2. effective relative damage coefficients for omni-directional space electrons and protons of various energies for solar cells with various cover-glass thicknesses;
3. Space radiation environment data for the orbit of interest.

The equivalent 10 MeV proton fluence can be converted to equivalent 1 MeV electron fluence as follows: $\Phi_{1MeV\ e} = 3000\ \Phi i_{10MeV\ p}$.

In cases when the cell degradation is entirely dominated by proton damage, the cell degradation could be estimated more accurately by calculating the equivalent 10 MeV proton fluence and using 10 MeV proton cell damage data, than by using the equivalent 1 MeV electron fluence and electron data.

To use cover-glass darkening data, a procedure is necessary to evaluate the absorbed dose produced by the various radiation components of the space environment. The procedure is similar to that used for equivalent fluence, with the exception that the absorbed dose varies with depth in the cover material.

6. The power and energy budget

The starting point for the solar array sizing is the correct identification of the power demand throughout the whole mission of the spacecraft.

Such power demand may change during the satellite lifetime either because of different operational modes foreseen during the mission or, more simply, because of degradation of the electrical performances of the electrical loads (in majority electronic units).

Taking into consideration what just said, an analysis of power demand is performed, including peak power, of all the loads installed either in the platform or as payload for each identified phase of the mission. Because of presence of sun eclipses, and possible depointings along the orbit, an analysis of the energy demand is also performed, this because in case of insufficient illumination the on board battery will supply the electrical power, and the solar array has to be sized in order to provide also the necessary power for its recharge. The power budget is based on peak power demands of the loads, while the energy budget is based on average consumptions.

It is good practice consider power margins both at unit and electrical system level.

The consumption of each unit is calculated considering the following criteria:

• 20% margin with respect to expected power demand if the unit design is new.
• 10% margin if the unit design has a heritage from a previous similar one.
• 5% margin if the unit is recurrent.

Several electronic units work in cold or hot redundancy; this has to be taken into account when summing the power demands.

Once the power demand is defined including the margins above, it is advisable to add 20% extra margin at system level and defined at the beginning of the project. Such margin is particularly useful during the satellite development in order to manage eventual power excesses of some units beyond the margins defined at unit level. In this way eventual Request For Deviation (RFD) issued by the subcontractors can be successfully processed without endangering the whole spacecraft design. This is particularly true for scientific missions, where many times the development of the instruments may reveal so challenging that an excess of power demand cannot be excluded a priori.

At this point harness distribution losses are introduced, 2% of the power demand defined with all margins at unit and system level may be a good compromise between losses containment and harness mass.

The Power Control and Distribution Unit (PCDU) is the electronic unit devoted for the solar array and battery power conditioning and regulation, power distribution and protection, execution of received telecommands (i.e. switch on/off of the loads) and telemetry generation. Its power consumption without considering the efficiencies of primary bus power converters depends on the management of the digital interfaces with the on-board computers, the control loop and protection electronics, the value of such consumption is not immediate to calculate but it can be said that a PCDU capable to manage 1kW can consume about 30W. However it consumption strongly depend on the number of implemented distribution lines, and relevant electronic protections.

Now its time to add the power needed for the recharge of the battery, this power strongly depend on the mission profile, and many times the maximum discharge of the battery occurs at launch, from lift-off up to the successful sun acquisition by the satellite with optimal sun pointing of the solar panels. Some times due to the complexity of the satellite design and mission profile it is not possible to have a full recharge of the battery in one orbit before the next eclipse, then the power allocated for such incumbency has to assure a positive battery recharge trend throughout a limited number of orbits.

The power delivered by the solar array is conditioned by suitable power converters in order to provide it to the loads with a regulated voltage, or at list with the voltage varying between a maximum and minimum value. These converters may have an efficiency between 98.5% and 95% and the choice of their topology is made according to several criteria and constraint dictated by the overall satellite system design. Such efficiencies are taken into account adding up to an additional 5% to the budget defined so far.

The harness losses between solar array and PCDU may be calculated having as objective 1V voltage drop at the maximum required power; again, considerations about the harness mass can provoke the change of such objective.

Finally, in case of the European ECSS standard (ECSS-E-ST-20C) is considered as applicable, an additional 5% margin on power availability shall be assured at the satellite acceptance review End of Life (EOL) conditions and one solar array string failed.

7. Solar array sizing; impact of the power conditioning and electromagnetic constraints

The definition of the solar array, conceived as a set of solar cells connected in series to form a string and strings connected in parallel cannot be made without considering the power conditioning device placed at its output in order to have the electrical power delivered within a certain voltage range. This is not the suitable seat for a complete examination of all the possible power conditioning and power architecture solutions, what can be said is that there are two main concepts: the Direct Energy Transfer (DET) and the Maximum Peak Power Tracking (MPPT). These two methods of regulation have an important impact on the solar array design not only from the sizing point of view, but also from the electromagnetic compatibility (EMC) one. The following section will detail the impact of the adopted power conditioning concept, and some sizing constraints mainly raised by the space environment such as electrostatic discharges and earth magnetic field.

7.1 Regulation based on Sequential Switching Shunt Regulator (S³R)

The first concept is based on the use of a shunt regulator; the figure below shows the electric schematic of a cell of a Sequential Switching Shunt Regulator (S³R), several solar array strings can be connected in parallel to the input of the regulator's cell; the voltage at the terminals of the output capacitor (Main Bus capacitor) is regulated by the switching of the MOSFET contained in the blue oval.

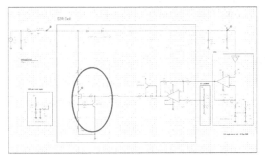

Fig. 10. Electrical Section of a Sequential Switching Shunt Regulator (S³R)

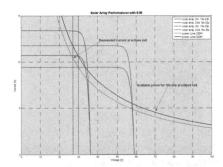

Fig. 11. Solar array working points as function of required power

The operating voltage of the solar array is constant and equal to main bus nominal voltage plus the voltage drops due the two diodes in series along the line, the solar array harness, and the blocking diode placed at the string positive output. In case of a fully regulated power bus, this operating voltage remains fixed during both sunlight and eclipse periods throughout the orbit; if the power bus is instead a battery regulated one it implies that the bus voltage decreases during eclipse periods, when the battery discharges, provoking a migration of the operating point of the solar array towards the short circuit one.

Supposing a power need of 280W, Figure 11 shows that a solar array composed of 20 strings of 18 cells (18s – 20p), at the eclipse exit (V_{array}= 27V) cannot provide the required power. In this condition the battery keeps discharging, lowering further down the operating voltage. This power bus lock-up has to be avoided increasing the number of strings in parallel. Adding 5 more strings (i.e. 25% more) the solar array can deliver 320W at 27V when cold;

therefore 40W become available to assure the battery charge. However, this increase might not be enough for assuring a full recharge of the battery in one orbit, or a positive recharge trend through several orbits; and an assessment of the energy budget by numerical simulation becomes necessary, taking into account orbital and attitude constraints.

7.2 Regulation based on Maximum Peak Power Point Tracker (MPPT)

The MPPT concept is based on the use of a switching dc-dc converter; usually it has a buck topology, where the primary voltage at solar array side is always higher of the secondary one on the distribution bus. Figure 12 shows an example of this type of converter. There are three control loops; a conductance control of the output current, an output voltage controller, and the Maximum Peak Power Tracker which regulates the output voltage of the solar array around the maximum power point in case of maximum power demand. In all the cases the required power is lower than the maximum available one the operating voltage of the solar array is kept between the maximum power voltage and the open circuit one.

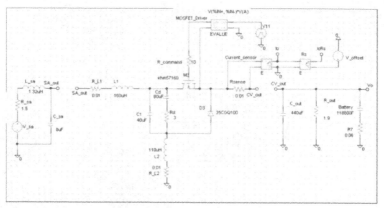

Fig. 12. Low ripple Buck converter topology

When this power conditioning concept is applied the solar array operating voltage is always independent from the bus one. Hence the phenomenon of the lock-up mentioned for the S3R is not present and the solar array does not need to be sized in order to cope with such issue.

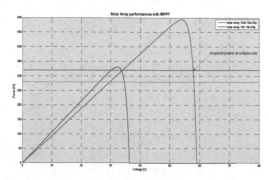

Fig. 13. Solar array P-V curves and required power, MPPT power conditioning

Figure 13 clearly shows that the original array composed of 20 strings is now capable to deliver the needed power in both hot and cold conditions, providing power to the loads (280W) and the additional 40W for the recharge of the battery.

Clearly from the sizing point of view of the array, the MPPT provides unquestionable benefits, but the price to be paid consist in additional mass (inductances and capacitances, as it can be seen in figure 12), and higher complexity because of the presence of three control loops.

7.3 Electromagnetic Compatibility (EMC)

The design of a spacecraft solar array and its power conditioner has to satisfy several requirements, not only in terms of mass, dimensions and power output, but also in terms of electromagnetic compatibility. This is particularly true for scientific mission, when instruments highly sensitive to electromagnetic fields may be boarded. In these cases it becomes crucial for the success of the mission to know which electromagnetic fields are generated at solar array level due to the circulating current and its frequency content, once this is connected to the power conditioning unit. The wires connecting the solar array to the PCDU, via the Solar Array Driving Mechanism (SADM) when necessary, are always twisted pairs (positive and return), but the return connections of the strings are routed on the rear side of the panel, they are not twisted of course, hence the solar array can behave as a transmitting antenna at frequencies which may result incompatible with some of the equipments on board.

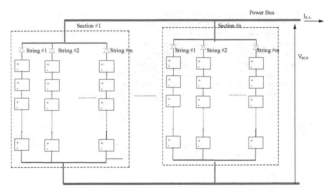

Fig. 14. Solar array electrical scheme

These issues are strongly dependent on the power conditioning approach adopted.

In the case of the S3R, with reference to figure 10, it can be seen that within the blue oval there is the shunt switch (MOSFET) together with a linear regulator in order to limit the current spikes at the regulator input when the MOSFET switches ON/OFF. Such spikes are strongly dependent on the total output capacitance of the strings connected in parallel and hence from the capacitance of the single triple junction solar cell. Fewer cells are in a string, or more strings in parallel, higher is this capacitance. The linear regulator can reduce the amplitude of the spikes by a suitable sizing of the dump resistor. For sake of completeness, the inductances present in the circuit diagram are the parasitic ones. Figure 15 shows the frequency spectrum of the current circulating in the harness between solar array and power regulator for different values of the dump resistor. The next figure 16 instead shows the frequency spectrum of the current for the same solar array section when the power conditioning is made by a buck converter with a MPPT control loop. It can be immediately seen that in case of MPPT power conditioning the current ripple on the solar array harness is much lower at low frequencies, not higher than 8 mA; and therefore such solution may be interesting when the power subsystem has to cope to very stringent requirements from EMC point of view.

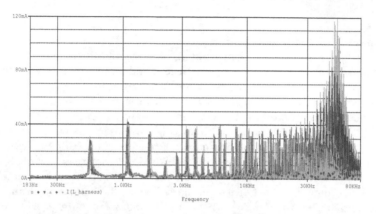

Fig. 15. Frequency spectrum of Solar Array output current for S³R power conditioning

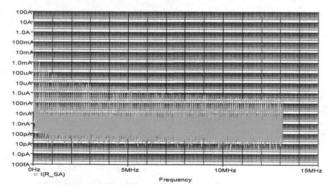

Fig. 16. Frequency Spectrum of Solar Array output current for MPPT power conditioning

7.4 Effect of the Earth magnetic field
The interaction between the Earth magnetic field B and the currents circulating in each string generate a torque disturbing the desired attitude of the whole spacecraft. The magnetic moment M due to the current is given by

$$M = I \cdot A \tag{16}$$

Where I is the current and A is the area of the current loop; in the case of the solar array this area corresponds in a first approximation to cross section of the panel substrate; on the front face of it the cells are mounted, on the rear face the return harness is implemented.
The resulting torque is

$$T = M \times B = M \cdot B \cdot \sin \vartheta \tag{17}$$

The direction of the torque is such that the dipole tends to orient itself parallel to lines of force of B, minimizing the potential energy and achieving a stable position.

This torque has to be in principle neutralised by the Attitude and Orbit Control System of the satellite, which implies the usage of thrusters (i.e. fuel consumption) or increased authority of magneto-torques and/or reaction wheels (electrical power and mass impact).

Clearly there are two ways for the minimisation of this torque; the first one is the minimisation of the areas of the current loops; the second one concern the layout of the solar array strings; adjacent strings can be disposed on the panels in opposite directions, such that the individual torques generated are balanced. With this solution, solar cells having the positive terminal at the string open circuit voltage will lay very close to cells having the negative terminal at 0V. And this opens the door to another issue to be faced.

7.5 Electrostatic Discharges (ESD)

The space plasma is the cause of the accumulation of electrostatic charges on the spacecraft surfaces. The energy of the plasma changes with the altitude; it is around 10,000 eV at about 36,000 km (Geostationary Orbits, GEO) decreasing to 0.1 eV for below 1,000 km (Low Earth Orbits, LEO), within the Van Allen Belts. For what concern the solar arrays it can be said that the interconnections between solar cells and the cell edges are exposed to plasma, and the output voltage resulting at the terminals of a string plays an important role. The worst scenario occurs at BOL, at the minimum operative temperature (eclipse exit). In these conditions the open circuit voltage is at the maximum value, if triple junction solar cells are used and a string is for instance composed of 34 cells, this voltage can be above 90V; this is the maximum voltage between two adjacent cells.

The value of the maximum current that can flow through a conductive part of the array (usually the current of a single string if each is protected by a diode) is also important; indeed it has been proofed that in order to have a self sustained secondary arc, minimum value of the current for a particular voltage is needed. In case of ECSS standard applies, in particular "Spacecraft Charging – Environment Induced Effects on the Electrostatic Behaviour of Space Systems (ECSS-E-20-06)", then it can be said that no tests are required to prove the safety of the solar array to secondary arcing when the maximum voltage-current couple available between two adjacent cells on the panel, separated with 0.9mm as nominal value, is below the threshold in the following table:

VOLTAGE	CURRENT	COMMENTS
70 V	0.6 A	No self sustained secondary arcing possible
50 V	1.5 A	No self sustained secondary arcing possible
30 V	2 A	No self sustained secondary arcing possible
10 V	-	Voltage is too low to allow any arcing

Table 2. ESD limit conditions

An inter-cell gap between strings of adjacent sections may be defined at 2 mm, cell to cell, that means 1.85 mm between cover-glasses. Finally, taking into account tolerances of the tools used during manufacturing of the solar array, it results that the distance between adjacent strings is always higher than 1.6 mm

8. Solar array configurations

The solar arrays mounted on a satellite can have very different shapes, accommodations and dimensions. The configuration of a solar panel is the result of several design iterations made at satellite level, considering the mission requirements, the needed power, the dimensions, mass, and the spacecraft attitude to be kept during the whole lifetime and in all the possible satellite working modes. However three or four main configurations of the solar array can be identified.

8.1 Spinning satellite

The first configuration is the one characterising a spinning satellite. The satellite usually has cylindrical shape with the symmetry axis as the rotation one. This configuration was the first one to be adopted; the available power is not elevated with respect to the panel surface, indeed the equivalent active area results from the division of the actual area of panel by π.
The satellite Meteosat is a good example; this configuration is nowadays rarely used, but in some cases is still interesting for scientific satellites like those of the Cluster mission.

Fig. 17. Solar Array for spinning Satellite, Meteosat Second Generation (Credits: ESA - MSG Team)

8.2 Body mounted panels

The second configuration foresees the panels body mounted to the spacecraft walls. The panels are rigidly fixed to the structure and their orientation towards to the sun is never optimal.

Fig. 18. Body mounted solar array, GOCE (Credits: ESA - AOES Medialab)

This solution has been recently adopted for earth observation and scientific satellites with a reduced power need, no more than 1 kW. In case of earth observation satellites the nadir-pointing attitude of the instruments results in highly variable illumination of the panel, therefore the computation energy budget can be quite challenging because the power subsystem may have power coming from both solar array and battery pack at the same time along the orbit. This behaviour may significantly reduce the useful time for the recharge of the battery in sunlight, and an oversized solar panel may be needed. The ESA spacecraft GOCE is a good example of such body mounted panels; two of them are installed on the fixed "wings" of the satellite, the other two are on the "fuselage". It is worth to note that the temperatures on the solar panels are very different between one another, this because of the different illumination levels and different thermal exchange of the wings (remaining colder) with respect to the fuselage (hotter panels). Such configuration, dictated by many other requirements at satellite level, can have a huge impact in the complexity of the power conditioning concept to be adopted.

8.3 Deployable wings

The third one is the classical double deployable wing. This solution is classical for telecommunication geostationary satellites. Each wing is moved by a Solar Array Driving Mechanism having the rotation axis perpendicular to the orbital plane. The illumination is optimized by the automatic orientation of the panels. This kind of configuration is the best solution when several kilowatts are needed, as in the case of recent telecom satellites. Each wing is then composed of several panels kept folded at launch, and then progressively deployed by suitable mechanisms at early phase of the mission. The satellite Hylas-1 gives a good example for such solution.

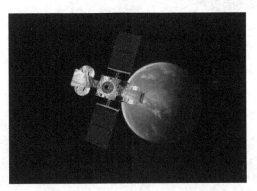

Fig. 19. Deployable Solar panels, Hylas-1 (Credits: ESA - J. Huart.)

9. Design and simulation examples

The following two examples will show how a spacecraft solar array, composed of one or more panels having different orientations, provide the needed power during the mission. The examples reported consider body mounted panels, having a fixed orientation with respect to the satellite body axes. This kind of panels is typically used for small and medium satellites, with a power demand less than 1 kW. If on one hand they are relatively cheap and easy to realise, on the other they may require additional efforts for the proper assessment of the energy budget throughout the orbits. This is particularly true in case of the power bus is an unregulated one, having wide voltage variations because of the battery charge and discharge cycling. In conclusion their design may be particularly challenging because of their typical small size, many times conditioned by the allowed dimensions and mass of the satellite, and the irregular illumination along the orbit.

The first example concerns an Earth observation satellite made as a cube. Three lateral sides are covered by solar cells; the fourth one accommodates the instruments and is Nadir pointing. The last two sides of the cube are parallel to the orbital plane. This configuration of the satellite is such that the illumination of each panel results to be almost sinusoidal, when the sun-light is incident on the panel itself. The temperatures will follow the same type of law, and the available electrical power as well. The orbit is sun-synchronous, and the transmitters are working when the ground stations are visible. The satellite is small; its required power is about 160W, and 60W are consumed by two different transmitters at different transmission frequencies. Each panel accommodates 8 strings of 18 cells each; the power conditioning is based on the S3R regulator with a battery power bus (battery directly connected to the distribution bus. This architecture is the one which can be prone to the lock-up of the power bus previously described, due to over-discharge of the battery after eclipse periods. The problem is however mitigated by the possibility to have a sun-bathing mode when the satellite passes over the oceans and in any case in the southern hemisphere. In this operative mode two of the three panels will have the common edge oriented towards the Sun, the sunlight incidence will be 45 deg. Figure 20 shows when these sun-bathing phases can occur (red ground-track).

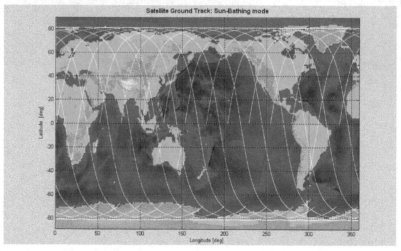

Fig. 20. Satellite ground track

As said, the required power is mainly function of the duty cycle of the transmitters when the ground stations are visible. In this example the three ground stations, typically used for earth observation missions are Kiruna (light blue), Fairbanks (magenta), and Redu (yellow). Figure 21 shows when these stations are visible, together with the eclipse periods (blue ground track). The illuminations of the panels for 24 hours (14 orbits) simulation are reported in figure 22. It can be clearly seen when the sun bathing occurs: panel #3 shows a constant illumination of about 950 W, while the panel #2 (magenta) has a slight increase due to the albedo effect; the panel #1 results to be not illuminated.

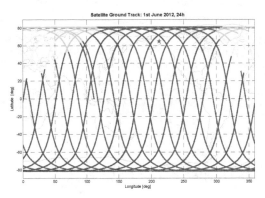

Fig. 21. Ground station visibility and eclipses

Figure 23 shows the calculated temperatures for the three panels. Finally, figure 24 reports the available power from the array, the power exchanged by the battery, and the power required by the loads; from this plot it can be clearly seen the power delivered by the battery is adequately balanced by the power used for recharging them.

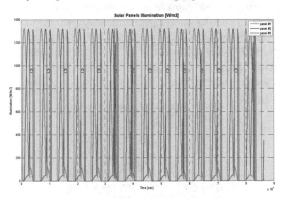

Fig. 22. Illumination of solar panels over 24 hours period

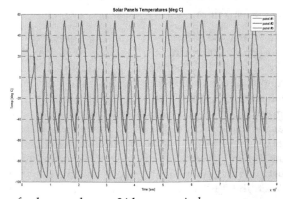

Fig. 23. Temperature of solar panels over 24 hours period

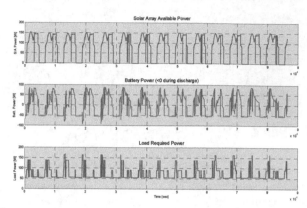

Fig. 24. Power Balance

The second example concern the design of a body mounted solar array which output power is conditioned by a MPPT control system. This is the case of LISA Pathfinder, which solar array is composed of 39 strings of 24 cells each, for 650W required power in EOL conditions. The nominal attitude during the mission is sun pointing, and the limited surface available for the solar array is due to mission and spacecraft configuration constraints. At a certain stage of the project it was decided to separate the solar panel from the rest of the structure by dedicated supports. This solution introduced the possibility to have different working temperatures between the strings and cells belonging to the same string, because of different thermal exchange modalities among centre and periphery of the panel.

Fig. 25. LISA Pathfinder artistic impression

Therefore it was worth to analyse whether a temperature gradient of 30 °C between centre and periphery may originate knees in the I-V curve that may be recognised as false maximum power points by the MPPT control loop, leading to a block of the working point of the array in a non optimal position. Figure 26 shows the layout of the solar cells within their strings, adjacent rows of cells of the same colour belong to the same string. The resulting I-V curve (green) of the whole array is showed in figure 27, as term of comparison the two V-I curves calculated considering constant temperature are also reported as term of comparison. To be observed that the cell with the lowest temperature in a string rules the maximum current flowing trough the string itself. From the plot it can be concluded there are no knees between the open circuit voltage and the maximum power knee such to provoke the lock of the MPPT tracker around a false maximum power working condition.

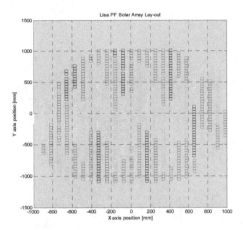

Fig. 26. Solar array layout

Figure 28 shows the illumination and the temperature reached by the solar panel in the first orbits after launch, the temperature over the panel is now considered as constant. It can be observed that the illumination takes into account also the contribution of the albedo just before and after an eclipse (no illumination), as expected from a solar panel always pointing towards to the sun throughout the orbit.

The figure 29 shows now the extended temperature profile over a period of 24 hours, together with output voltage and current; to be observed that from the fourth orbit onwards the temperature shows an slight increase after 70% of sunlight period has elapsed; this happens because when the battery is fully charged; the maximum power is not required anymore, the operating voltage of the array shifts toward the open circuit value. At the same time it can be seen that the output current decreases. This temperature increase is due to the difference between the maximum available power and the required one; the unused power warms up the array.

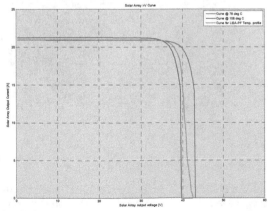

Fig. 27. LISA Pathfinder Solar array, V-I curve

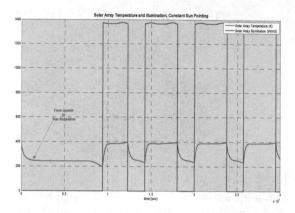

Fig. 28. Solar Array Illumination and temperature, launch phase and first 3 orbits

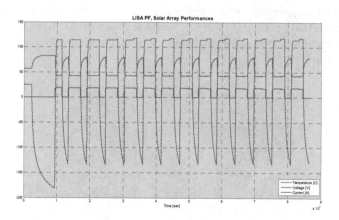

Fig. 29. Solar array temperature, output voltage and current

Finally, figure 30 shows the Depth Of Discharge (DOD %) of the battery from launch. The DOD is progressively recovered the first four orbits. After the fourth one, a stable charge-discharge cycling is reached.

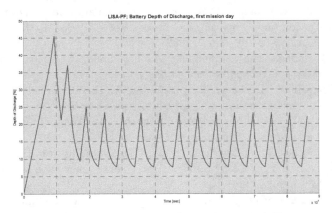

Fig. 30. Battery Depth of Discharge (DOD %) for launch phase and first mission day.

10. Conclusions

Objective of this chapter was to provide guidelines for the design at system level of a solar array for satellites. Such kind of application has to be compliant with severe requirements mainly dictated by the harsh space environment mainly in terms of temperature levels, cosmic radiations which provoke wide variations of the performances together with their continuous degradation. Mass and size of the panels are main constraints with respect to the required power as well as optimal orientation towards to the sun, several times limited by other requirements at spacecraft and mission level. The actual state of the art is represented by triple junction solar cells capable to have a bulk efficiency of more than 30%.

Typical accommodations of these arrays have been illustrated and a few design examples provided. These examples have been chosen among those may be considered as particularly challenging with respect to the required power and energy budgets coupled with mission constraints.

11. References

AZUR SPACE Solar Power GmbH, 3G-28% Solar Cell Data-sheet
        http://azurspace.de/index.php?mm=89

Strobl, G. et al.; (2002). Advanced GaInP/Ga(In)As/Ge Triple junction Space Solar Cells, *Proceedings of ESPC 2002 6th European Space Power Conference*, ESA-SP 502, Oporto, Portugal, May 2002.

Neugnot, N. et al.; (2008). Advanced Dynamic Modelling of Multi-junction Gallium Arsenide Solar Arrays, *Proceedings of ESPC 2008 8th European Space Power Conference*, Konstanz, Germany, Sept. 2008.

Tada, H. and Carter, J., Solar Cell Radiation Handbook, *JPL Report* 77-56, Caltech, Pasadena, 1977

Mottet, S., Solar Cells Modelisation for Generator Computer Aided Design and Irradiation Degradation, *ESA Symposium on Photovoltaic Generators in Space*, pagg. 1-10, Heidelberg, 1980.

Ferrante, J., Cornett, J. & Leblanc, P., Power System Simulation for Low Orbit Space craft: the EBLOS Computer Program, *ESA Journal* Vol 6, 319-337, 1982.

Diffuse Surfaces, *ESA PSS-03-108* Issue 1, 1989

O'Sullivan, A. Weinberg: The Sequential Switching Shunt Regulator (S[3]R); *Proceedings Spacecraft Power Conditioning Seminar,* ESA SP-126, 1977

Colombo, G., Grasselli, U., De Luca, A., Spizzichino, A., Falzini, S.; Satellite Power System Simulation, *Acta Astronautica,* Vol. 40, No. 1, pp 41-49, 1997.

De Luca, A. et al.; The LISA Pathfinder Power System, *Proceedings of ESPC 2008 8th European Space Power Conference,* Konstanz, Germany, Sept. 2008.

De Luca, A., Chirulli, G.; Solar Array power Conditioning for a spinning satellite, *Proceedings of ESPC 2008 8th European Space Power Conference,* Konstanz, Germany, Sept. 2008.

De Luca. A.; Simulation of the Power System of a Satellite, graduation thesis, *ESA EAD (European Aerospace Database),*Quest Accession Number 96U03072, 1996. or *Database NASA,* Quest Accession Number 96N48163, 1996.

9

Power Output Characteristics of Transparent a-Si BiPV Window Module

Jongho Yoon

Hanbat National University
Republic of Korea

1. Introduction

Energy-related concerns about traditional resources include the depletion of fossil fuel, a dramatic increase in oil prices, the global warming effect caused by pollutant emissions from conventional energy resources, and the increase in the energy demand. These concerns have resulted in the recent remarkable growth of renewable energy industries [1-3]. Furthermore, renewable energy has become a significantly important research area for many researchers as well as for governments of many countries as they attempt to ensure the safety, long-term capability, and sustainability of the use of global alternative energy resources [2]. Renewable energy resources include solar, geothermal, wind, biomass, ocean, and hydroelectric energy. [4] In particular, both solar (i.e. photovoltaics) and wind energy are considered to be leading technologies with respect to electrical power generation.

The study of photovoltaics (PV) has been carried out since the 1980s' and is currently the most significant renewable energy resources available. According to the Renewable Energy Policy Network for the 21st Century (REN21), there has been a strong growth in the use of PV of 55 % and the worldwide solar PV electric capacity is expected to increase from 1,000 MW in 2000 to 140,000 MW by 2030 [5]. Moreover, it is forecast by the European Renewable Energy Council that this renewable electric energy could become sufficient to cover the base load and half of the global electricity energy demand by 2040 [6]. Generally in the PV industry, crystalline silicon has generally occupied about 95 % of the market share of materials, while only 5 % of all solar cells use amorphous silicon [7]. However, in order to improve the cost efficiency of solar cells by using less material, the thin-film PV module with amorphous silicon has become an active research and development (R&D) area [8]. In particular, solar cells that use amorphous silicon have the advantage of being able to generate a higher energy output under high temperatures than crystalline silicon solar cells, which are less affected by the temperature increase with respect to performance of electricity output than are the crystalline silicon solar cells. Moreover, installed at the rooftop and on the exterior wall of the building, a thin-film solar cell can be conveniently used as a façade that generates power for the entire building. This system is known as a building integrated photovoltaic system (BIPV). The thin-film solar cell can also provide the advantage of heat insulation and shading when incorporated into a harmonious building design. Therefore, the thin-film solar cell is expected to be a very bright prospect as a new engine for economical growth in the near future. Currently in Korea, many researchers are conducting vigorous research on PV with respect to the application of crystalline silicon solar cells. An example of such research includes the evaluation of the power output of PV modules with respect to the ventilation of the rear side of the module. However, research on the transparent thin-film solar cell as a building façade application including windows and doors is only in its early stages.

Therefore, the objective of this study is to establish building application data for the replacement of conventional building materials with thin-film solar cells. In this study, an evaluation is carried out on the performance of the thin-film solar cell through long-term monitoring of the power output according to the inclined slope (the incidence angle). This is conducted by using a full-scale mock-up model of the thin-film solar cell applied to a double glazed system. In addition, the aim of the application data of the thin-film solar cell is to analyze the effect of both the inclined slope and the azimuth angle on the power output performance by comparing this data with the simulation data for PV modules[9].

2. Methodology

In this study, a full-scale mock-up model was constructed in order to evaluate the power output performance of a PV module laminated with a transparent thin-film solar cell. A mock-up model was designed for a PV module that had a range of inclined slopes, and was used to measure the power output according to the slope (incidence angle) and the azimuth angle. The collected experimental data was then compared with the simulated data for a power performance analysis.

A commercialized single plate transparent thin-film solar cell with amorphous silicon was used in this study (KANEKA, Japan). This was modified into a double glazed PV module in order to install the mock-up model for this study.

Using the full-scale mock-up model, the system output was monitored for 9 months. A computer simulation (TRNSYS, University of Wisconsin, USA) of the PV module was also performed at the same time, and empirical application data was calibrated for the statistical analysis of power performance based on the inclined slope and the azimuth angle. In particular, the annual power output of the PV module was obtained by analyzing the data obtained from the remaining 3 months on the basis of the 30 years' standard weather data in Korea.

3. Double-glazed PV module

In Korea, it is an obligatory requirement that building materials such as windows and doors for a residence should be double glazed in order to ensure adequate heat insulation. Moreover, as the demand for energy efficiency buildings increases, the efficiency of double glazed window systems is improving with respect to heat insulation, as is the efficiency of exterior wall systems of buildings. Therefore, the photovoltaic characteristic of thin-film solar cells was measured in terms of the transmittance of the cell prior to evaluation of the PV module (Figure 1). The results of this measurement showed an average transmittance of 10 % at the range of visible radiation between 390 nm and 750 nm.

Using this thin-film solar cell, a single plate PV module was manufactured to a thickness of 10 mm, and the PV module was then modified as a double glazed module of 27 mm thick, consisting of a 12 mm air space and a 5 mm thick layer of common transparent glass, as shown in Figure 2.

J.-H. Song et al. / Energy and Buildings 40 (2008) 2067–2075

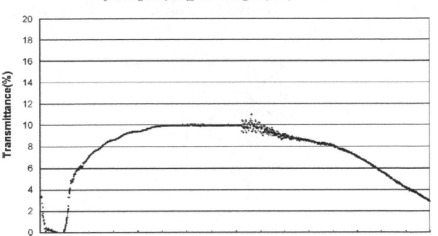

Fig. 1. Transmittance of PV module depending on the wavelength

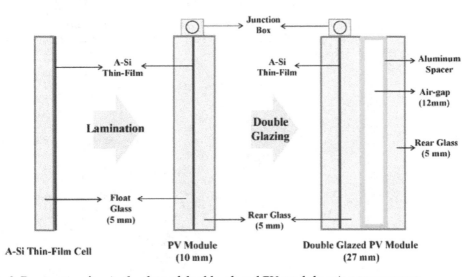

Fig. 2. Preparation for single plate of double-glazed PV module using transparent amorphous silicon (A-Si) thin-film cell.

From the performance evaluation of the heat insulation, the prepared PV module exhibited a 2.64 W/m²-℃ thermal transmittance, as shown in Figure 3. However, it showed an 18 % solar heat gain coefficient (SHGC), which was much lower than that measured for the common double glazed window. WINDOW 6.0 and THERM5.0 (LBNL, USA) were used to analyze the heat insulation of the standard type of double glazed PV module widely used for the heat insulation of building windows and doors. This analysis allowed for the evaluation of heat transfer under a two dimensional steady state for the user defined fitting system at a given circumstance.

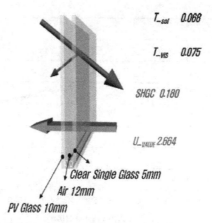

T_{-sol} 0.068

T_{-vis} 0.075

SHGC 0.180

U_{-VALUE} 2.664

Clear Single Glass 5mm
Air 12mm
PV Glass 10mm

Fig. 3. Optical and thermal characteristics of double-glazed PV module (T_sol is the solar transmittance, T_vis is the transmittance of visible radiation, SHGC is the solar heat gain coefficient, and U_value is the thermal transmittance of PV module).module

Figure 4 shows a plane figure of a 10 mm thick and 980 × 950 mm single plate PV module, and a PV module consists of 108 cells in series. The electrical characteristics of the prepared PV module are listed in Table 1.

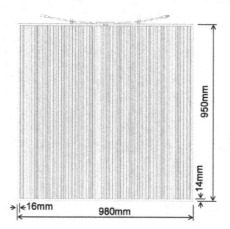

950mm

14mm

16mm   980mm

Fig. 4. Plane figure of a single plate PV module.

4. Full-scale mock-up model

A full-scale mock-up model was constructed with the dimensions of 8 m long, 5 m wide, and 3.5 m high, as shown in Figure 5. In order to demonstrate the impact of the inclined

Item	Specification
Module thickness (mm)	10
Module efficiency (%)	7
Maximum power output (W)	44.0
Maximum voltage (V)	59.6
Maximum electric current (A)	0.74
Open circuit voltage (V)	91.8
Short circuit current (A)	0.972

Table 1. Specification of the tested thin-film PV module

slope (incidence angle) on the power output, the inclined angles were varied on the mock-up by installing both a tilted roof at 30° and a common roof without any slope. The mock-up faced south in order to maintain a compatible solar irradiance with the location of Yongin, Gyeonggi, Korea. Two separated spaces were prepared in order to test the thin-film PV module (Test room A in Figure 5(a)) and the common double glazed window (Test room B in Figure 5(a)) as a reference. The spaces were 2 m long, 3 m wide, and 2.7 m high. The double glazed PV module and the common double-glazed window were installed in each separated test room at different inclined angles (0 °, 30 °, and 90 °).

A mock-up model was also constructed in order to monitor the electric current, voltage, power, temperature, and solar irradiation depending on the inclined angle of the PV module. The double glazed thin-film PV module revealed only a 10 % transmittance (See Figure 1), but this was as sufficient as the common double glazed window for observing the outside.

5. Power performance of PV module

5.1 PV module performance measured in mock-up model
The total solar irradiance and power output of the PV module, depending on the inclined angle of double glazing, were monitored through the mock-up model for 9 months from November 2006 to August 2007. Data obtained from the mock-up was collected based on minute-averaged data, and the final data of 12,254,312 was statistically analyzed based on 56 variables. Firstly, daily data was rearranged into monthly data. Secondly, minute-based data was averaged and combined into an hourly data. Finally, each group was analyzed in terms of an arithmetic mean, standard deviation, minimum, and maximum value. The empirical data in this study was limited in DC output, which was obtained from the load using resistance without an inverter. Thus, it is assumed that there may be a number of differences between the data measured in this study and the empirical data controlled by maximum power peak tracking (MPPT) using an inverter.

Figure 6 shows the hourly data, which was yearly-averaged, of the intensity of solar irradiance and DC output depending on the inclined angle of the double glazed PV module. Based on the data measured at noon, the inclined slope of 30 ° (SLOPE _30) revealed an insolation of 528.4 W/m^2, which shows a greater solar irradiation than that for the slopes of 0 ° (SLOPE_0, 459.6 W/m^2) and 90 ° (SLOPE_90, 385.0 W/m^2), as shown in Figure 6(a). Consequently, the average power output at noon also exhibited 19.9 W for SLOPE_30, which was higher than that shown in the data for SLOPE_0 (15.76 W) and SLOPE_90 (8.6 W) (See Figure 6(b)).

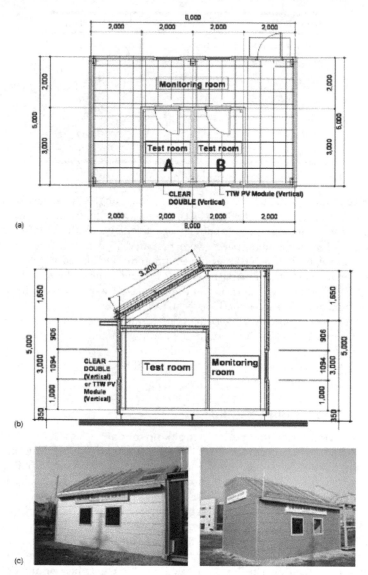

Fig. 5. Full-scale mock-up model: (a) a floor plan view, (b) a cross-sectional view, and (c) photographs of mock-up model.

5.2 Effect of intensity of solar irradiance

Figure 7 depicts the relationship between the solar irradiance taken from the PV module and the DC power output depending on the inclined angle of the module. For all PV modules, the power output increased with an increase in solar irradiance. While the increase rate of power output was particularly retarded under the lower solar irradiance, there was a very steep increase of power output under the higher solar irradiance (See Figure 7).

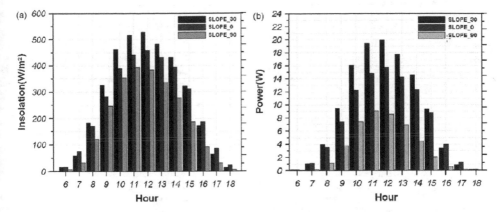

Fig. 6. Monitoring data of PV module depending on the slope through the mock-up model: hourly data averaged yearly: (a) solar irradiance and (b) power output.

By observing the degree of scattering for each inclined PV module as shown in Figure 7, there was a higher density of power output distribution for SLOPE_30 under the higher solar irradiance. On the other hand, the lowest distribution of power output was revealed for SLOPE_90, even under the higher solar irradiance. The monthly-based analysis revealed that a double glazed PV module inclined at 30 ° (SLOPE_30) produced the greatest power output due to the acquisition of a higher solar irradiation. This result can also be achieved from a PV module with an incidence angle of 40.2 °, implying that it is more efficient to acquire solar irradiation than any other factor (See Figure 7(b)).

In the case of SLOPE_0, there were significant differences in power output with respect to solar irradiance depending on monthly variation (See Figure 7(a)). Specifically, the maximum solar irradiance in December is only 500 W/m² resulting in a power output of 10 W. On the other hand, the maximum solar irradiance of 1,000 W/m² with over 50 W power output was recorded for June. This high efficiency of power performance for SLOPE_0 during the summer could be due to the incidence angle of 36.1 °, which was low enough to absorb solar irradiation.

The reverse tendency of power output for SLOPE_0 was shown for SLOPE_90, which was installed at the horizontal plane. Specifically, a maximum power output of above 30 W was observed. This was due to a quiet efficient solar irradiance with the maximum solar irradiation gain of over 900 W/m² occurring in December. However, a lower solar irradiance of around 500 W/m² with less than 10 W power output was observed during the summer months from June to August. This can be explained by the difference in the incidence angle of the PV module depending on the inclined slope, i.e., the lower incidence angle of 36.6 ° for SLOPE_90 was observed during the winter, particularly in January, while the higher value of 84.6 ° was observed during the summer, especially in June. This implies that solar irradiation capable of producing a much higher power output can be easier to be achieved with a lower incidence angle of solar radiation to the PV module.

5.3 Monthly based analysis of power performance

Figure 8 shows the amount of solar irradiation and power output accumulated for each month depending on the inclined angle of the PV module. A fairly effective solar irradiance

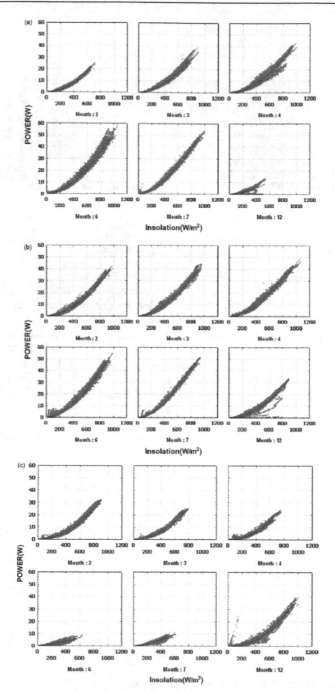

Fig. 7. Power output data of PV modules based on monthly variation of solar irradiance measured in the mock-up model: (a) slope 0, (b) slope 30, and (c) slope 90, respectively.

of 147.7 kWh/m² was obtained from SLOPE_30 during May, and the lowest value of 75.3 kWh/m² was obtained in December (See Figure 8(a)). The horizontal module of SLOPE_0 resulted in the highest solar irradiance in June and the lowest value in January. On the other hand, the PV module installed at the vertical window exhibited the highest solar irradiance (115 kWh/m²) in January and the lowest (50.2 kWh/m²) in August. This can be explained by the highly effective solar irradiance of both of the PV modules that were installed horizontally (SLOPE_0) and tilted at a slope of 30 °. This was due to the smaller incidence angle, defined as the angle between the incident solar ray and the normal line, close to the horizontal plane during the summer and related to the height of the sun, while the PV module installed vertically (SLOPE_90) obtained an effective solar irradiance due to the smaller incidence angle during the winter.

An analysis was also carried out on the monthly power performance depending on the inclined angle of the PV module, as shown in Figure 8(b). From the monthly data in Figure 8(b), it can be seen that the most effective power output during the summer, particularly for June, was obtained at SLOPE_30 and SLOPE_0. However, the highest power output was obtained at SLOPE_90 for January. This could be due to the variation of solar irradiance from each PV module from the different incidence angles based on the height of the sun.

In this study, the best power performance among all the tested PV modules was that obtained by the PV module tilted at an angle of 30 ° (SLOPE_30), comparing with those installed horizontally (SLOPE_0) and vertically (SLOPE_90).

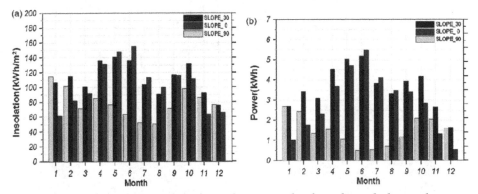

Fig. 8. Monitoring data of PV module depending upon the slope through the mockup model: monthly accumulated data of (a) solar irradiance and (b) power output.

5.4 Hourly based analysis of power performance

Figures 10~12 show the statistically analyzed monthly power generation data of PV module depending the inclined slope. The name of each part is provided for the better understanding in Figure 9. ■ sign in each box indicates Mean value, □ and ▨ signs indicate the range of Mean±S.D (Standard Deviation), Whisker I sign indicates the range between maximum and minimum values. For example, in the first graph of Figure 10, the mean value at 12pm in January is approximately 20W, S.D. (Standard Deviation) is 5~30W, maximum value is 40W and the minimum value is 0W. The statistical data on how much power is generated in each hour can be easily understood with these graphs. Furthermore, the maximum and minimum ranges can also be easily analyzed, enabling the comparison of characteristic behaviors depending on the inclined angle.

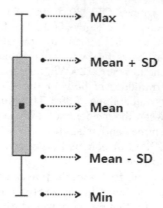

Fig. 9. Explanation of Box-Whisker graph

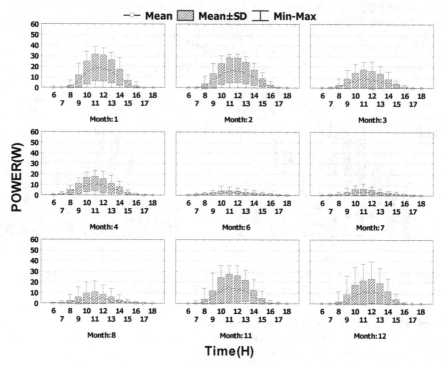

Fig. 10. Power generation of SLOPE._90° in each timestep

In case of vertical PV module, the power generation turns out be significant in Janurary due to a farily effective solar irradiance. It showed the power generation of 20W on average at noon. On the other hand, in June when there is no high solar irradiance due to high incidence angle, the power generation was less then 10W on average at noon. The inclined slope of 30 ° showed the best power generation during the measurement period. Especially the power generation was the greatest in June with 30W on average at noon.

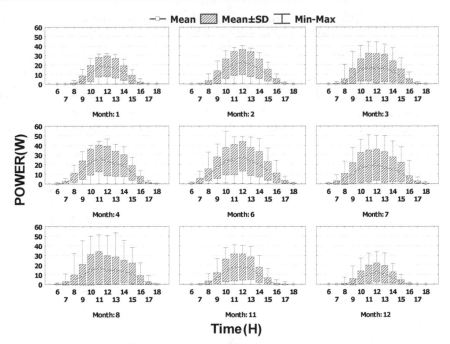

Fig. 11. Power generation of SLOPE._30° in each timestep

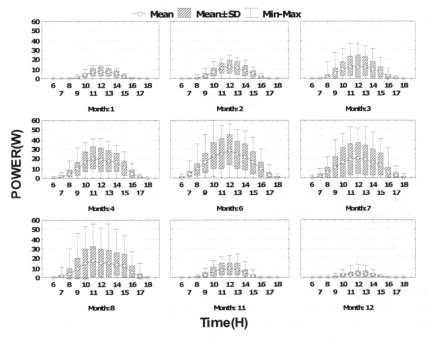

Fig. 12. Power generation of SLOPE._0° in each timestep

In case of horizontal PV module, it showed effective power generation performance in the summer similar to the case of the inclined slope of 30 °, showing more than 30W generation on average at noon. However, the generation barely exceeded 10W in December due to high incidence angle and low solar irradiance. The hourly average power generation depending on each inclined angle is illustrated in Figure 13. In case of inclined angle (SLOPE_30), it showed power generation of 20W on average at noon, while the horizontal PV module showed 15W on average. Vertical PV (SLOPE_90) showed the low generation performance of 8W on average. Table 5.5 summarizes the hourly average power generation, voltage and electric current.

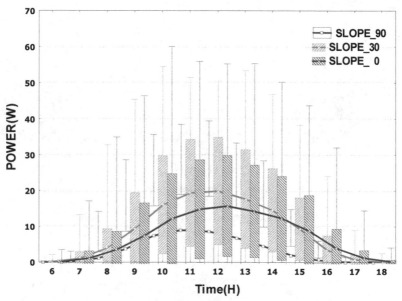

Fig. 13. Annual hourly averaged power generation

5.5 Analysis of power performance through simulation

In this study, TRNSYS (Ver. 14.2, Solar Energy Laboratory, Univ. of Wisconsin, USA) was used as a simulation program to analyze the performance of power output for a double glazed PV module. Generally, TRNSYS has been widely used to compute the hourly data for power output, solar irradiance, temperature, and wind speed for both PV systems and solar heat energy systems [10]. From the simulation program, the relative error was verified, and a comparison was then made of the power output from the experimental and the computed data, as shown in Figure 14. In addition, the experimental data from the PV module with an inclined angle of 30 ° (SLOPE_30) was compared with the simulated data in terms of the annual power output: 1,060 kWh/kWp was obtained from the experiment and 977 kWh/kWp was estimated from the computational simulation. This computed data showed a relative error of 8.5 %, which is considered to be a reliable result within the error tolerance. Thus, the computational simulation was conducted to demonstrate the power output performance of a PV module installed at various inclined angles.

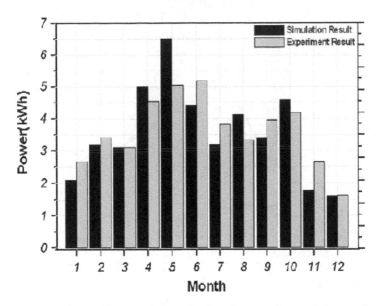

Fig. 14. Power output data calibration by comparing the experimental data to the computed data obtained from the simulation program (TRNSYS).

Power performance analyses were performed of PV modules facing south (azimuth = 0 °) depending on the different inclined angles of 0 °, 10 °, 30 °, 50 °, 70 °, and 90 °. The data set consisted of the experimental data for 0 °, 30 °, and 90 ° and the computed data for 10 °, 50 °, and 70 °. Figure 15 illustrates the monthly power output depending on the inclined angle ranging from 0 ° to 90 ° south (azimuth = 0 °). PV modules that were tilted at an angle below 30 ° showed a relatively good power performance of over 6 kWh in the summer, while those with an inclined angle above 50 ° demonstrated a power performance of less than 6 kWh. The most effective annual power output data of 977 kWh/kWp was obtained at an inclined angle of 30 ° (SLOPE_30), as shown in Figure 16. On the other hand, the lowest annual power output of 357 kWh/kWp was obtained from the PV module with a slope of 90 ° (SLOPE_90), which was 37 % of the annual power output of SLOPE_30. From Figure 16, it can be seen that the annual power output performance was effective in the order of SLOPE_10 (954 kWh/kWp), SLOPE_0 (890 kWh/kWp), SLOPE_50 (860 kWh/kWp), and SLOPE_70 (633 kWh/kWp).

The power generation performance depending on the angle of the azimuth was also estimated for PV modules with different inclined slopes, as shown in Figure 17. Similarly, a PV module inclined at an angle of 30 ° showed the most effective power output data for all directions in terms of azimuth angles, and the lowest data was obtained from that with an inclined angle of 90 °. For the PV module inclined at an angle of 30 °, the best power performance among the analyzed PV modules facing various directions was obtained for the PV module that was installed to the south (azimuth = 0 °). It can be seen from Figure 17 that different azimuth angles affected the power performance of PV modules: that is, the power performance decreased as the direction of the PV module was changed from the south to the east and west, in comparison to the PV modules that were inclined at the slope of 30 °, as listed in Table 2.

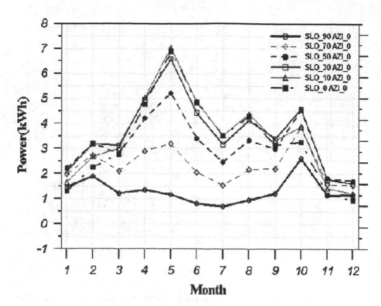

Fig. 15. Monthly power output data of PV module depending on the slope, and facing south (azimuth = 0).

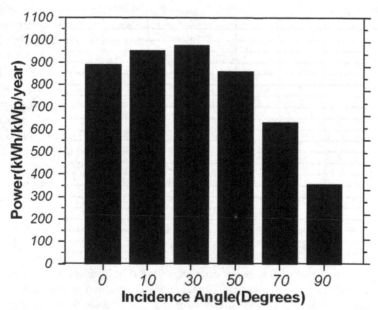

Fig. 16. Annual power production of PV module depending on the slope, and facing south (azimuth = 0).

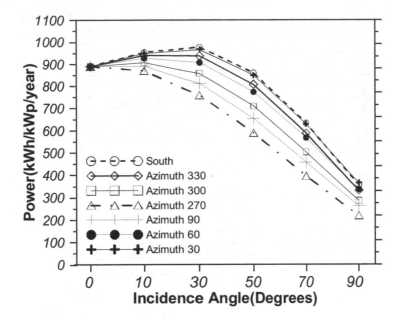

Fig. 17. Annual power production of PV modules with various slopes depending on the angle of azimuth ranging from 0 to 90

Angle of azimuth (°)	Direction	Power performance efficiency[a] (%)
0	South	100
30	Southwest 30 °	99
60	Southwest 60 °	93
90	West	83
270	East	78
300	Southeast 60 °	88
330	Southeast 30 °	96

a. Power performance efficiency was calculated from the percent of power output at each azimuth angle on the basis of the power output data of PV module to the south.

Table 2. Power performance efficiency of PV module with a slope of 308 depending on azimuth angle

It can be seen from Figure 17 that for the annual power performance of several PV modules, the power output increased with an increase of the inclined angle below 30 °, and decreased with an increase of the inclined angle above 30 °. In particular, at inclined slopes above 60 ° there was a steep decline of power performance with the increase of the inclined slope, as shown in Figure 17. This could be due to the incidence angle modifier correlation (IAM) of glass attached to the PV module, which showed a similar tendency in IAM depending on the inclined angle [11], as can be seen in Figure 18. Actually, IAM should be computed as a function of incidence angle (θ) when estimating the power output of the PV module, by using the following Equation (1) [11]:

$$\text{IAM} = 1 - (1.098 \times 10^{-4})\theta - (6.267 \times 10^{-6})\theta^2 + (6.583 \times 10^{-7})\theta^3 - (1.4272 \times 10^{-8})\theta^4 \qquad (1)$$

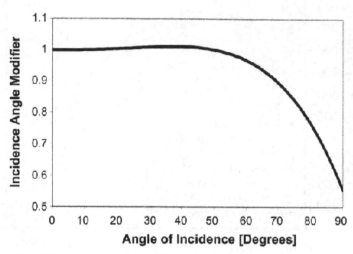

Fig. 18. Correlation of incidence angle modifier given by King et al. (1994).

Accordingly, a characteristic of the glass attached to the PV module is considerably influential so that the solar transmittance (Tsol) remarkably decreases with an increase in the inclined slope of the PV module from the higher incidence angle. Therefore, the solar transmittance efficiency can significantly affect the power output of the PV module.

6. Power efficiency of PV module

6.1 Hourly based analysis of the power efficiency
The power efficiency can be calculated by multiplying total irradiation by the PV window area. Annual averaged power efficiency is illustrated in Fig. 19.

$$\eta_{S,\tau} = \frac{E_{use,\tau}}{A_a \text{ X } H_\tau}$$

$\eta_{S,\tau}$; Power Efficiency
$E_{use,\tau}$; Power Output(Wh)
A_a ; PV windows area (m²)
H_τ ; Total irradiation on the PV windows
Annual average power efficiencies of the inclined slope of 30 ° (SLOPE_30), horizontal PV module (SLOPE_0) and vertical PV module (SLOPE_90) turned out to be 3.19%, 2.61% and 1.77%, respectively, indicating that the inclined slope of 30 ° showed the greatest efficiency. On the other hand, the horizontal PV showed the highest instantaneous peak power efficiency of 6.0% followed by those of the inclined slope of 30 ° (5.6%) and vertical PV (4.0%) angles. In terms of the monthly average power efficiency depending on each inclination angle, the inclined slope of 30 ° (SLOPE_30) showed 3.82% in June and the horizontal PV (SLOPE_0) showed 3.63% in July. The inclined slope of 30 ° showed 2.15 % of efficiency and the horizontal PV showed 0.81% in December. On the other hand, the vertical

PV (SLOPE_90) showed the peak efficiency of 2.38% in February and lowest efficiency of 0.80% in June. The inclined slope of 30 ° (SLOPE_30) showed the greatest annual average power efficiency of 3.19%, followed by horizontal and vertical PV modules showing efficiencies of 2.61% and 1.77%, respectively.

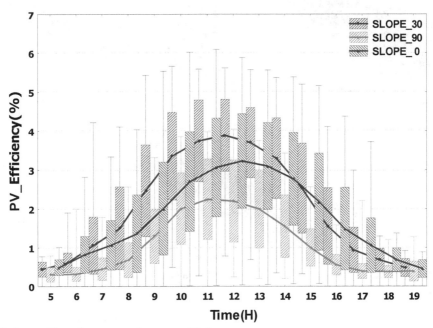

Fig. 19. Annual hourly averaged power efficiency

6.2 Effect of power efficiency by the intensity of solar irradiance

Assuming the solar irradiance of 900 W/m², the power efficiencies of the inclined slope of 30° and horizontal PV reached 5%, while the vertical PV partially exceeded 3%. The inclined slope of 30 ° and horizontal PV showed relatively high power efficiency even under high solar irradiance conditions, while the efficiency of vertical PV significantly dropped after reaching 500W/m². The inclined slope of 30 ° and horizontal PV can obtain relatively uniform solar irradiance throughout the year and thus the high power efficiency can be achieved over the large range of solar irradiance, while the vertical PV absorb the low solar irradiance during the winter period and thus the power efficiency is reduced in those low irradiance conditions.

6.3 Power efficiency by the temperature variation

The correlation between the power efficiency and the PV surface temperature variation is illustrated. Under the low solar irradiance, the data is scattered and thus did not show the clear correlation. However, it showed the clear correlation between PV efficiency and the surface temperature under the solar irradiance higher than 600W/m², i.e., the PV efficiency is improved at higher surface temperature. This is due to the fact that the higher surface temperature enhances the power efficiency in case of amorphous PV as opposed to crystalline silicon solar cell (c-Si solar cell).

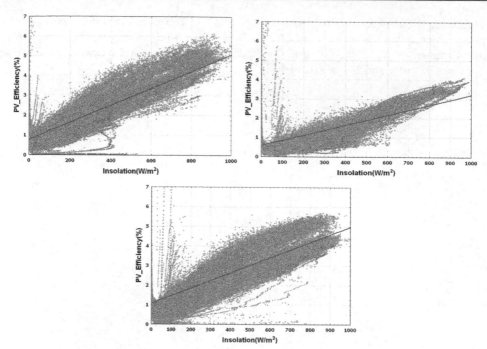

Fig. 20. Correlation between solar insolation and power efficiency (SLOPE_90°, SLOPE_30°, SLOPE_0°)

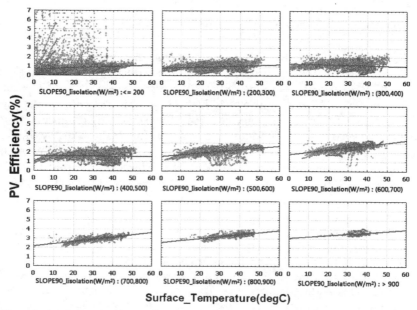

Fig. 21. Correlation between the surface temperature and power efficiency (SLOPE_90°)

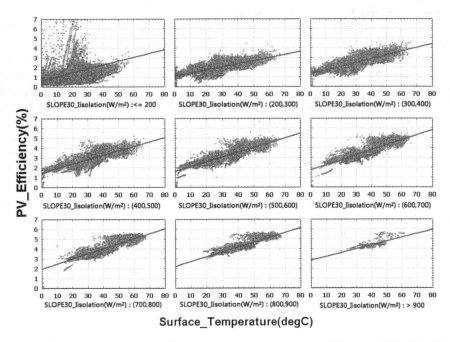

Fig. 22. Correlation between the surface temperature and power efficiency (SLOPE_30°)

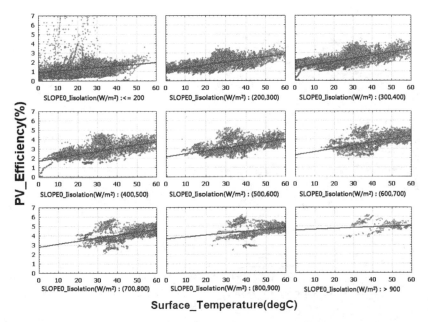

Fig. 23. Correlation between the surface temperature and power efficiency (SLOPE_0°)

6.4 Power efficiency by the solar incidence angle

The PV efficiencies of each inclination angle under different solar incidence angle and solar irradiance are illustrated in the figures below. In case of vertical PV module (SLOPE_90), the power efficiency showed constant value until the solar incidence angle of 65° and it started to rapidly drop after 65°. These characteristics are considered to be the effect of absorbed solar insolation (incident angle modifier) depending on the solar incidence angle reaching the PV module glass wall. This phenomenon did not take place in case of the inclined slope of 30 ° (SLOPE_30) due to the low PV efficiency at the solar incidence angle higher than 65°. Likewise, the horizontal PV module was not affected by incident angle modifier as well in most of the solar radiation conditions except for the high solar incidence angle of greater than 65° and the low solar insolation of less than 400W/m² where the efficiency was rather decreased.

It turns out that the power efficiency of PV module is largely affected by the solar incidence angle, solar azimuth and altitude. Furthermore, the rapid decrease in the PV efficiency during the summer period is due to the reduced solar transmittance through the window system at the solar incidence angle higher than 70°, showing the impact of the front glass of PV module on the power efficiency.

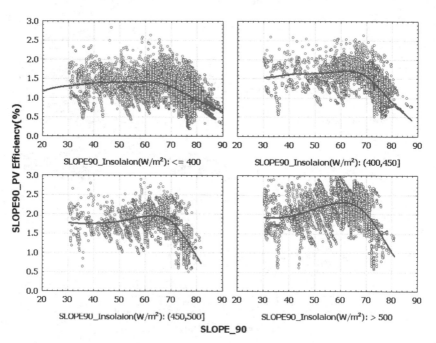

Fig. 24. PV module power efficiency vs. solar incidence angle (SLOPE_90°)

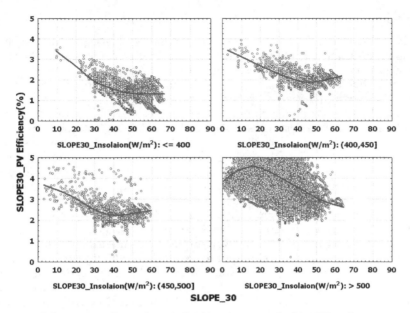

Fig. 25. PV module power efficiency vs. solar incidence angle (SLOPE_30°)

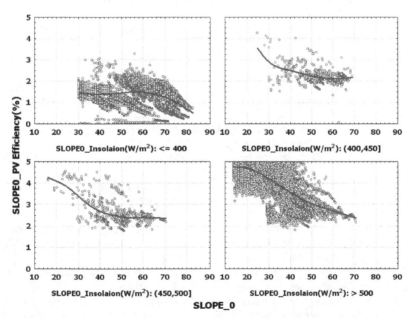

Fig. 26. PV module power efficiency vs. solar incidence angle (SLOPE_0°)

7. Conclusion

This study evaluated a transparent PV module in terms of power generation performance depending on installation conditions such as the inclined slope (incidence angle) and the azimuth angle. The objective of this evaluation was to provide useful data for the replacement of traditional building windows by BIPV system, through the experimental results measured in the full-scale mock-up system.

1. The annual power output of the PV module was measured through the mock-up model. The PV module that was installed at a slope of 30 ° exhibited a better performance of 844.4 kWh/kWp annual power output than the vertical PV module with a slope of 90 °.
2. The experimental data was compared with the computed data obtained from the simulation program. The computed data is considered to be reliable with a relative error of 8.5 %. The best performance of annual power output was obtained from the PV module with a slope of 30 ° facing south, at an azimuth angle of 0 °. The inclined angle was one of the factors that significantly influenced the power generation performance of the PV module, which varied within a range of 24 % on average and provided a maximum difference of 63% in the power output at the same azimuth angle.
3. In terms of the computed power output from a slope of 30 ° depending on the azimuth angle, the PV module facing south exhibited the most effective performance compared to other azimuth angles. The direction in which the PV module faces can also be a very important factor that can affect the power performance efficiency by 11 % on average and by a maximum of 22 %, depending on the azimuth angle.

8. References

[1] Y. Kuwano, Progress of photovoltaic system for houses and buildings in Japan, Renewable Energy 15 (1998) 535–540.
[2] A. Ja¨ger-Waldau, Photovoltaics and renewable energies in Europe, Renewable and Sustainable Energy Reviews 11 (2007) 1414–1437.
[3] A. Stoppato, Life cycle assessment of photovoltaic electricity generation, Energy 33 (2008) 224–232.
[4] A. Hepbasli, A key review on exergetic analysis and assessment of renewable energy resources for a sustainable future, Renewable and Sustainable Energy Reviews 12 (2008) 593–661.
[5] A. Zahedi, Solar photovoltaic (PV) energy; latest developments in the building integrated and hybrid PV systems, Renewable Energy 31 (2006) 711–718.
[6] S. Teske, A. Zervos, O. Schafer, Energy revolution, Greenpeace International, European Renewable Energy Council (EREC) (2007).
[7] R.W. Miles, G. Zoppi, I. Forbes, Inorganic photovoltaic cells, Materials Today 10 (2007) 20–27.
[8] S. Guha, Amorphous silicon alloy photovoltaic technology and applications, Renewable Energy 15 (1998) 189–194.
[9] J.H. Song, Y.S. An, S.G. Kim, S.J. Lee, Jong-Ho Yoon, Y.K. Choung, Power output analysis of transparent thin-film module in building integrated photovoltaic system(BIPV), Energy and Building, Volume 40, Issue 11, (2008) 2067-2075
[10] TRNSYS, A transient system simulation program version 14.2 Manual. Solar Energy Laboratory: University of Wisconsin, Madison, USA, 2000.
[11] D.L. King, et al., Measuring the solar spectral and angle of incidence effects on photovoltaic modules and irradiance sensors, in: Proceedings of the IEEE Photovoltaic Specialists Conference, 1994, pp. 1113–1116.

10

Innovative Elastic Thin-Film Solar Cell Structures

Maciej Sibiński and Katarzyna Znajdek
*Technical University of Łódź, Department of
Semiconductor and Optoelectronic Devices,
Poland*

1. Introduction

The idea of thin films dates back to the inception of photovoltaics in the early sixties. It is an idea based on achieving truly low-cost photovoltaics appropriate for mass production, where usage of inexpensive active materials is essential. Since the photovoltaic (PV) modules deliver relatively little electric power in comparison with combustion-based energy sources, solar cells must be cheap to produce energy that can be competitive. Thin films are considered to be the answer to that low-cost requirement [1].

Replacement of single crystalline silicon with poly and amorphous films, caused the decline of material requirements, which has led to lower final prices [2]. Furthermore, the thickness of cell layers was reduced several times throughout the usage of materials with higher optical absorption coefficients. Unique, thin film and lightweight, devices of low manufacturing costs and high flexibility, were obtained by applying special materials and production techniques, e.g. CIS, CIGS or CdTe/CdS technologies and organic elements. Taking advantage of those properties, there is a great potential of new, useful applications, such as building integrated photovoltaics (BIPV), portable elastic systems or clothing and smart textiles as well [3].

Low material utilization, mass production and integrated module fabrication are basic advantages of thin film solar cells over their monocrystalline counterparts [4]. Figure 1 (by NREL) shows the development of thin film photovoltaic cells since 1975.

The development of cadmium telluride (CdTe) based thin film solar cells started in 1972 with 6% efficient CdS/CdTe [5] to reach the present peak efficiency of 16.5% obtained by NREL researchers in 2002 [6]. Chalcopyrite based laboratory cells (CIS, CIGS) have recently achieved a record efficiency of 20% [7], which is the highest among thin film PV cells (see Table 1). Solar modules based on chalcopyrites, uniquely combines advantages of thin film technology with the efficiency and stability of conventional crystalline silicon cells [4].

Thin film solar cell type	CIGS	CdTe/CdS	a-Si
Cell area [cm²]	0.5	1.0	0.25
Highest efficiency [%]	20.0	16.5	13.3
Typical efficiency range [%]	12.0 – 20.0	10.0 – 16.5	8.0 – 13.3

Table 1. Efficiencies of CIGS, CdTe and a-Si thin film solar cells [8].

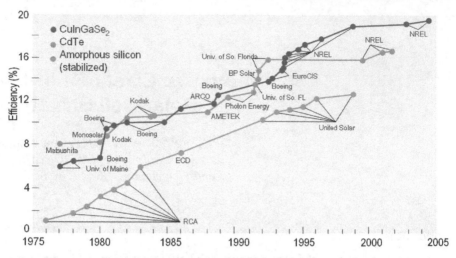

Fig. 1. Best laboratory PV cell efficiencies for thin films [source: www.nrel.gov]

To comprehend the developmental issues of thin films, it is important to examine each individually. Each has a unique set of advantages and shortcomings in terms of their potential to reach the needed performance, reliability and cost goals [1].

1.1 CdTe cells and modules

Cadmium telluride based solar cells are one of the most promising in thin film photovoltaics. With a bandgap of 1.45 eV the material well suited to match the AM 1.5 solar spectrum. Furthermore, its high absorption coefficient causes that only a few microns absorber film is required for solar cell operation. The typical thin film CdTe/CdS structure is shown on Figure 2. Figure 3 presents the total life-cycle Cd emissions to prove that CdTe based PV cell are environment friendly and health safe.

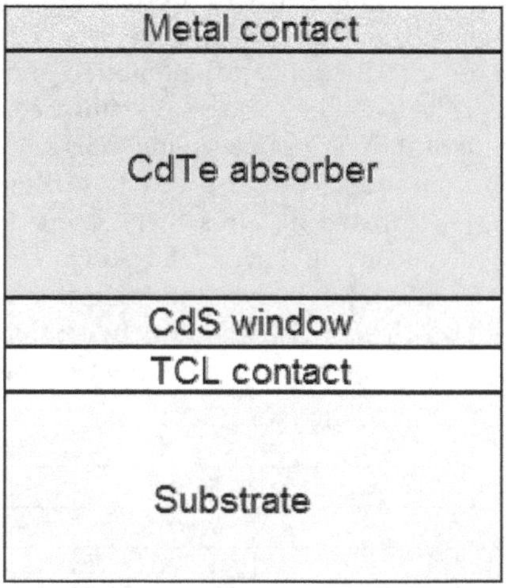

Metal contact
CdTe absorber
CdS window
TCL contact
Substrate

Fig. 2. Typical structure of CdTe thin film solar cell.

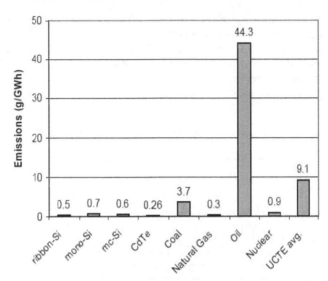

Fig. 3. Total life-cycle Cd emissions by Brookhaven National Laboratory [9].

Low-cost soda-lime glass, foil or polymer film can be used as the substrate of CdTe/CdS solar cell. The best results of 16.5% efficiency are achieved with glass substrate (Table 1). However, Laboratory for Thin Films and Photovoltaics at EMPA, Switzerland obtained 12.7% efficiency of single CdTe solar cell on polymer foil and 7.5% of monolithically interconnected flexible CdTe solar module of 32 cm^2 total area [10]. Transparent conductive layers (TCL) are usually thin conductive metal oxides, such as ITO (*Indium Tin Oxide*), Zn_2O_4, Cd_2SnO_4, ZnO:Al, CdO, ZnO, In_2O_3, SnO_2 or $RuSiO_4$. However, lastly in flexible solar cells, transparent conductive oxides (TCO) are being replaced by carbon nanotube (CNT) composites [11] or graphene. The CdS film is grown either by chemical bath deposition (CBD), close space sublimation (CSS), chemical vapor deposition (CVD), sputtering or vapor transport deposition (VTD). For the growth of CdTe, three leading methods are used for module fabrication: CSS, electro-deposition (ED) and VTD. Wide variety of metals can be used as back contact for thin film CdTe solar cells, e.g. Cu, Au, Cu/Au, Ni, Ni/Al, Sb/Al, Sb/Au [12].

Several thin-film PV companies are actively involved in commercializing thin-film PV technologies. In the area of CdTe technology major players are or were in the past: First Solar (USA), Primestar Solar (USA), BP Solar (USA), Antec Solar (Germany), Calyxo (Germany), CTF Solar (Germany), Arendi (Italy), Abound Solar (USA), Matsushita Battery (Japan) [8, 12, 13]. This effort lead to 18% share of CdTe cells in global PV marked in 2009 [14].

1.2 CIS/ CIGS/ CIGSS structures

Other promising material for thin film solar cell absorber layer is copper indium diselenide $CuInSe_2$. CIS has a direct bandgap of 0.95 eV which can be increased by the addition of gallium to the absorber film. About 30% of Ga added to CIS layer (CIGS cell), changes the bandgap from 0.95 eV to almost 1.2 eV, which improves its match with the AM 1.5 solar spectrum. Higher gallium content (of 40%) has a detrimental effect on the device performance, because of its negative impact on the charge transport properties. The best gallium to indium ratio is 3:7 for high efficiency PV devices. The role of sulfur in CIGSS is to increase the bandgap of the absorber film [12], which can boost the AM 1.5 spectrum fitting even more. The typical thin film CIGS solar cell structure is shown on Figure 4. Figure 5 presents an example CIGS cell structure manufactured in the Laboratory for Thin Films and Photovoltaics at EMPA, Switzerland.

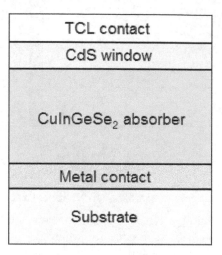

Fig. 4. Typical structure of CIGS thin film solar cell.

Fig. 5. CIGS cell structure manufactured at EMPA, Switzerland [8].

Worldwide, several companies are presently offering commercial thin-film PV CIGS products: Würth Solar, (Germany), Global Solar, (USA), Honda, (Japan), Showa Shell, (Japan), Sulfurcell, (Germany), Solibro (Germany), Avancis (Germany), Solyndra (USA), Centrotherm (Germany). Also, worldwide, about 34 companies are actively developing thin-film CIGS PV technologies. These companies are using different deposition methods for growing the thin CIGS absorber layers, as is shown in Table 2 [13].

Company	Substrate	Back contact	Process	Front contact
Shell Solar	Glass	Mo	Sputter/ Selenization	ZnO
Global Solar	Steel	Mo	Co-evaporation	ITO
Miasole	Glass	Mo	Sputter	ZnO
Wurt Solar	Glass	Mo	Co-evaporation	ZnO
Avancis	Glass	Mo	Sputter/ RTP	ZnO
Daystar Tech.	Glass	Mo	Sputter	ZnO
EPV	Glass	Mo	Sputter/ Evaporation	ZnO
Ascent Solar	Polymer	Mo	Co-evaporation	ZnO
ISET	Glass/Flexible	Mo	Ink/ Selenization	ZnO
Nanosolar	Flexible	Mo	Print/ RTP	ZnO
Solo Power	Steel	Mo	ED/ RTP	ZnO

Table 2. Thin film CIGS technology.

The absorber layer for commercial products uses either co-evaporation or the two-stage process such as the deposition of the precursors (Cu, Ga, In) by sputtering followed by selenization. Absorber layer can be made out of three chalcopyrite chemical compounds: $Cu(In,Ga)Se_2$, $Cu(In,Ga)(S,Se)_2$, $CuInS_2$ which are respectively CIGS, CIGSS and CIS.

As a back, metal contact Mo deposited by sputtering is most commonly used. Majority of companies (Table 2) use ZnO as the front transparent conductive layer. Zinc oxide is deposited either by sputtering or chemical vapor deposition [13]. Window layer of CdS (or alternatives, such as ZnS [8]) can be grown by analogous methods as in CdTe solar structure (CBD, CSS, CVD etc.).

As a substrate of thin film CIGS solar cell, either glass, metal (steel) sheet or polymer might be chosen. Highest efficiencies, as noted in Table 1, were achieved for modules on glass substrate. However, such solution have several inconveniences, which are for example: bulkiness, fragility and heaviness. Flexible substrates, on the other hand, offer both manufacturing and application related advantages, such as: large active area, roll-to-roll high speed (throughput) deposition, high material utilization, low thermal budget, monolithic interconnection, lower costs, light-weight and flexibility. Table 3 presents the comparison of thin film CIGS solar cell on steel and polymer substrate.

To conclude this subsection, we should ask a question: "why thin film CIGS and CdTe solar cells are worth attention?". The answer was given and it can be summarized in several points:

1. they are highly efficient,
2. active layers can be deposited on various substrates including flexible ones,
3. they have extremely stable performance,
4. they cause no environmental or health hazards,
5. they are low cost,
6. these cells are attractive for both terrestrial and space applications [8].

Metal (steel)	Polymer
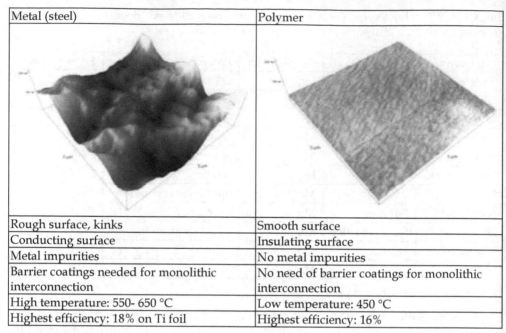	
Rough surface, kinks	Smooth surface
Conducting surface	Insulating surface
Metal impurities	No metal impurities
Barrier coatings needed for monolithic interconnection	No need of barrier coatings for monolithic interconnection
High temperature: 550- 650 °C	Low temperature: 450 °C
Highest efficiency: 18% on Ti foil	Highest efficiency: 16%

Table 3. Substrates for flexible CIGS solar modules [8].

Thin-film photovoltaic cells and modules are already widely popularized, mainly because of their small production costs and relatively high efficiency [15]. Moreover, some other, significant advantages, such as small weight and flexibility may be offered. That is the reason why large number of applications is being pursued using thin-film PV technologies, including building-integrated photovoltaics (BIPV), roof-top applications and utility-scale applications [13].

2. Manufacturing technologies of elastic thin – film CdS/CdTe solar cells

Cadmium telluride solar cells are placed among thin-film polycrystalline structures of high optical absorption and relatively low material consumption, with long development history [16]. Numerous virtues of each semiconductor compound as well as good cooperation within p-n heterojunction give the opportunity of efficient and cheap monolithic solar modules construction. Additionally, elastic structure of polycrystalline layers enables the flexibility of the manufactured modules and gives possibility of new implementations. The experiments leading to this goal were undertaken by various manufacturing technologies, however to maintain high optoelectronic parameters of the manufactured cells, independently on employed technology, proper polycrystalline structure must be preserved.

Standard technology of CdTe solar cells manufacturing is based of column grain structure.

This specific material organization occurs owing to vertical growth of hexagonal CdTe and CdS grains and gives the opportunity of high optical generation, smooth vertical charge carrier transport and thus high conversion efficiency. Both semiconductors are predestined

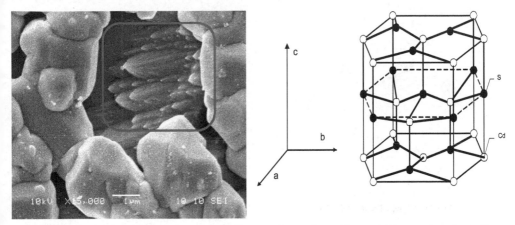

Fig. 6. SEM picture and the diagram of CdS wurtzite grains with vertical growth orientation and CdS hexagonal grain model.

to achieving of this structure under some technology circumstances [17] and may be matched with crystal constant differences not higher than 9,7% [18]. Structure of model CdS layer, obtained by authors, organized in wurtzite phase is presented in Figure 6.

The most popular manufacturing technologies of CdS/CdTe solar cells are nowadays CVD *(Chemical Vapour Deposition)* and the variants like PECVD *(ang: Plasma Enhanced CVD, or* MOCVD *(Metall Organic Chemical Vapour Deposition)* [19],CBD *(Chemical Bath Deposition) and physical methods like* PVD *(Physical Vapour Deposition),* CSS *(Close Space Sublimation)* [20], and variants of CSVT *(Close Space Vapour Transport)* [21, 22]. Alternatively screen-printing technology was also successfully employed for production of relatively thick CdTe base [23]. Morphology of the last mentioned layers was verified by authors with the help of SEM analysis indicating dense compact structure of hexagonal grains (Figure 7).

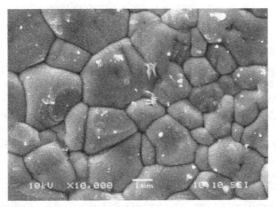

Fig. 7. SEM picture of dense, compact CdTe grain layer up to 8μm, manufactured by ICSVT technology on glass substrate.

As the additional experiments AFM profile of this layer, presented in Figure 8 was prepared.

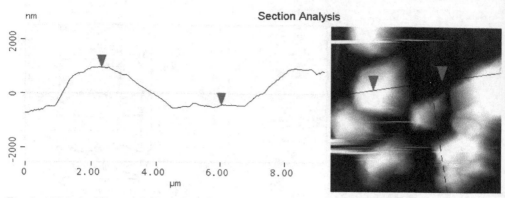

Fig. 8. AFM profiling of CdTe polycrystalline layer, made by ICSVT technology.

This measurement gives some important information about grains structure and inter-crystal surfaces. By means of polycrystalline CdS/CdTe layers profiling one may easily detect the diameter and grain shape but also the inter-grain valleys depth and possible structure fluctuations and layer discontinuities. These structure disorders may result in serious parameter losses by producing of shunt interconnections or other charge flow parasitic effects. Taking into account cadmium dichloride dissolvent presence, frequently caused by recrystallization demands morphology defects may cause a real and serious threat for CdS/CdTe structure functioning. No such phenomenon was confirmed by presented results.

All new production techniques of thin –film polycrystalline solar cells and particularly CdS/CdTe structure, designed for new application field, should be verified according to obtained layers profile to eliminate structure disorder. Since AFM spectroscopy gives the profiling results for some strictly limited area for wider statistical examination mechanical profiling of high accuracy may be applied. These experiments were also conducted for test CdS/CdTe structure. The experiment for each sample was conducted by the help of mechanical profilometer Dektak 3 VEECO Instruments. The measurements were performed for representative 100 μm scan range with 50nm resolution.

As the first analysis ICSVT CdTe layer morphology was checked (Figure 9). By this investigation the average grain diameter of 6-8μm was confirmed and the typical roughness of 1200 A. Series of measurements confirmed dense, compact structure of absorber grains with no interlayer shunts. Obtained average roughness and the CdTe grain surface profile suggests that inter-grain trenches are insufficient for significant degradation of shunt resistance value. The total layer level fluctuations are smaller than 4 μm, which taking into consideration typical glass thickness accuracy confirms homogenous thickness of the whole base layer.

Alternative technology, used for CdS/CdTe cell layers deposition is a standard PVD method. Under some circumstances it offers a possibility of semiconductor material deposition even on profiled, elastic and untypical substrates. Evaporation of examined materials caused serious technology problems connected with proper thickness of obtained layers and homogeneous structure of deposited material. Some serious parasitic effect like boiling, splitting and granulation of the material were solved by proper temperature profile and optimized one-directional tantalum evaporation source adopted by authors [24].

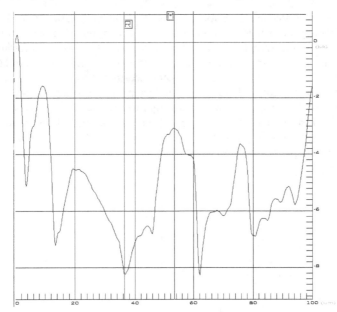

Fig. 9. Morphology of CdTe solar cell base absorber manufactured by ICSVT technology on glass substrate.

The investigation of CdTe base manufactured by evaporation and subsequent recrystallization is presented in Figure 10.

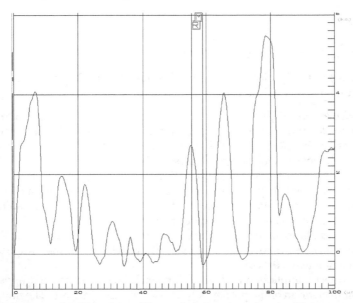

Fig. 10. Morphology of CdTe solar cell base absorber manufactured by evaporation and subsequent high-temp. recrystallization.

The measurements indicated homogenous grains of the average dimension 2-4µm. Typical roughness of 1230 µm is similar layers obtained by ICSVT, however one may observe higher peak values of grain top-trench profile. This may result in some interlayer parasitic connections. Additionally in this case, total layer thickness fluctuation is similar to the layer thickness (±2 µm), what may cause some absorber discontinuities.

Third investigated technology (Figure 11) is based on adaptation of screen-printing technique for semiconductor layers manufacturing. Printing paste is produced by milling of cadmium and tellurium in stochiometric proportion and mixing with special binder. After printing, leveling and drying process CdTe layer is recrystallized in high-temperature process similarly to the previously described way.

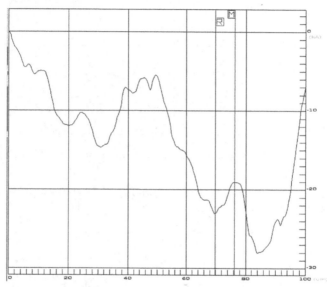

Fig. 11. Morphology of CdTe solar cell base absorber manufactured by screen-printing and subsequent high-temp. recrystalization.

Profilometry of screen-printed base (Figure 11) indicated very high thickness of investigated layer (even up to 20 µm), but also serious morphology defects. Typical roughness of obtained layers is beyond 3000µm, which exceeds 240% of average value for ICSVT and PVD techniques. Moreover obtained grains present different diameters from 1-2 µm up to 10 µm. Total layer thickness also varies strongly (locally up to 50%), which results in various optoelectronic parameters. Layers produced by screen-printing technology presents additionally high porosity, which prevents cadmium dichloride from volatilization and causes fast base oxidation and thus cell parameters degradation.

Considering the results obtained for three technologies aimed at manufacturing of CdS/CdTe solar structure on novel substrate material conclusion of their applicability and further development may be drawn.

ICSVT being the most complicated and so-far not commercialized method appeared to be the most efficient in creation of the proper polycrystalline base structure. The morphology of obtained layers confirms proper column grain structure of hexagonal CdTe crystals and thus high electrical parameters of final solar cell. Taking this into consideration further development of this technique is desired in terms of non-flat architectonic elements as the solar cell substrate. First steps towards this goal have been undertaken.

Evaporation and post annealing of CdTe material in order to formation of proper base structure appeared to be less efficient. However simple and effective for various substrates technique leads to thin film base production only. Moreover the diameters of single grains obtained by the author are significantly lower than by ICSVT technology.

Screen – printing and sintering as the last investigated method appeared to be insufficient for the stabile cell-base production. The advantage of high layer thickness is seriously diminished by poor homogeneity, thickness instability and high surface porosity. Nevertheless screen printed layers can be effectively used as the in-production material for ICSVT or similar processes.

3. Innovative polycrystalline elastic structures, based on polymer substrates

Although CdS/CdTe cells have now entered the mass - production phase, but still there are many possibilities of their new applications fields. Basing on this idea, authors proposed the implementation of modified CdS/CdTe cell structure in universal, attractive application called BIPV (Building Integrated Photovoltaics) and also elaborated elastic cell structure [25]. The CdTe cell construction gives the opportunity of achieving these goals, under the conditions of the proper technology modifications, as well as proper substrate and contacts implementation. Due to successful application of CSS variants of CdS/CdTe manufacturing technology for effective solar cells production, further experiments towards new cell structure and properties became possible. Considering cell composition, two opposite configuration of CdTe cell became possible. Historically first one is a classical substrate configuration (Figure 12 a), whereas based on glass + ITO, emitter-based configuration is called superstrate (Figure 12 b).

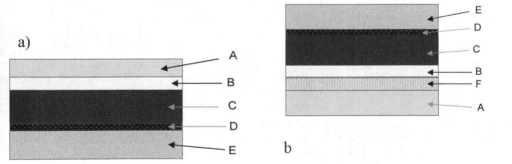

Fig. 12. Substrate a) and superstrate b) configuration of CdS/CdTe solar cell. A- glass cover, B- CdS emitter, C-CdTe base, D – base P+ sub layer, E-back contact, F-TCO layer, emitter metal contacts not visible.

Both of them possess some important advantages and technology drawbacks. Substrate configuration offers more mature manufacturing technology and lower substrate demands, while superstrate configuration ensures higher efficiencies (smaller surface shadowing) and better encapsulation. Adaptation of the described technology for new application and cell construction, demands deep consideration of all possible solutions.

Every introduced concept posses some value according to different aspects of BIPV applications and each is subsequently investigated by Technical University of Łódz research group. Ceramic substrates could be recognized as the best platform for the complete integration of the photovoltaic element with the architectonic component. One may find the reports on practical investigation of this construction for other thin –film solar cells e.g. CIS devices [26]. However, for CdS/CdTe construction, there is still research and technology adaptation needed. Additionally this kind of application is strictly connected with one particular architectonic element type such as roof-tile or brick. Moreover it has to provide the complete modular interconnection and regulation system, since the whole installation is made of hundreds of elements, working in different conditions. Furthermore, different interconnection systems (series, parallel and series-parallel) are necessary for optimum power and load polarization. Finally, standard ICSVT/CSS technology needs some fundamental modifications, in case of implementation in profiled architectonic elements (roof tiles or ornaments), since the material transport occurs only between very closely positioned source and target.

Taking into account cadmium telluride solar cells, possessing flexible construction two base materials may be considered. One is thin metal foil, while the second is the polymer material. Implementation of metal foils, for example Mo substrates, for CdTe construction has been already investigated and reported [27]. In this work we focus on polymer foil implementation as the elastic solar cell substrate. Flexibility of this material, combined with policrystalline thin-film structure properties, gives a promise that manufacturing of elastic solar panel, ready for integration with any shape architectonic substrate is possible. Moreover, it offers the opportunity of constructing both substrate and superstrate configuration of CdS/CdTe cell. Additionally, polymer foils are lightweight, high-durable materials, which enhances the possible application field of cells. Depending on the configuration, production technology and desired application different properties of the substrate foils will be demanded. Finding proper foil material and appropriate technology adaptation are the key to obtain efficient elastic PV cells.

To define the properties of polymer base foils, one may consider the specific of each configuration. So far, in the superstrate configuration highest conversion efficiencies were obtained However, in this case, polymer substrates must meet several conditions. One can mention as the most important: high optical transparency in the full conversion range of CdS/CdTe cell, ability of TCO surface electrode covering, high thermal durability, high chemical and water resistance. Apart from these specific demands, substrate foil of any configuration is expected to be light-weight, have high elongation coefficient, thermal expansion similar to semiconductor polycrystalline layers (CdS and CdTe) and be low cost. In both cases elastic cells can be easily attached to different shape architectonic elements.

Taking this into account, also substrate configuration of elastic cadmium telluride cell was investigated. As the preliminary step possible polymer material options were verified. Polymers, as the materials, are constructed on a base of multi-modular chains of single, repetitive units called monomers. In the manmade polymers, even the number of a few thousand monomer types is being achieved. The properties of manufactured polymer material depend strongly, not only on its chemical content and even monomer construction, but also on the monomers interconnecting system. Due to complexity of the typical polymer construction, it is impossible to evaluate the physical properties of these materials using theoretical analysis. This gave the prompt to the series of experiments, aimed at comprehensive evaluation of polymer foils physical parameters, potentially efficient as the CdS/CdTe cell substrate materials.

As the test group of polymer foils wide set of materials, including standard commercial solutions as well as high – temperature polyester and polyamide, was accepted. Among polyamide foils of high thermal durability, two materials - KAPTON® and UPILEX® foils were chosen. Both of them are commercially available high-technology materials implemented in specific applications (eg: space shuttles wings and nose cover, high power loudspeakers membranes). They are characterized by high mechanical and thermal durability, high dielectric constant and UV durability. Among the polyester materials high – temperature MYLAR® material was adopted. As the reference material, popular PET foil in standard and high - temperature production version was applied. First evaluation step of material properties is a verification of their mechanical parameters. Comparison of these results is presented in Table 2.

Parameter\Foil	PET/High temp PET	UPILEX®	MYLAR®	KAPTON® HN 100
Thickness [μm]	25.0	30.0	30.0	25.4
Weight [g/m²]	30.0	44.1	41.7	35.0
Surface mass coefficient [m²/kg]	31.2	22.7	23.98	27.9
Thermal expansion [%/ 1°C]	0.025	0.018	0.007	0.005
Standard elongation (25 °C) [%]	600.0	54.0	103.5	40.0

Table 2. Main mechanical parameters of tested polymer foils.

Obtained parameters suggest similar properties of all investigated materials. However, some important differences are evident. The most important is the value of the thermal expansion coefficient (TEC). In general, one may say that in the case of high-temperature materials the value of thermal expansion is lower. Exceptionally, in the case of UPILEX® the value of this parameter is close to standard PET foil. According to considered configuration, thermal expansion coefficient of substrate foil should be adjusted to the value of the semiconductor base or emitter and contact layer. In both cases of semiconductor materials (CdS, CdTe), the value of TEC is very low (at the level of $5 \cdot 10^{-4}$ [% / 1°C]), but the most typical metal contacts present TEC value higher by the order of magnitude.

The critical parameter in the standard re-crystallization process, as well as in the ICSVT, is a thermal durability of layer material. The maximum values of declared operational temperature for each investigated foil are: 130°C for Standard PET, 185°C for High-temp PET, 254°C for Polyester MYLAR®, 380°C and 430°C for Polyamide KAPTON® foil. Basing on the declared temperatures and considering the ICSVT temperature demands, two most durable foils were accepted for further investigations. As the subsequent step the weight loss of KAPTON® and UPILEX® in higher temperatures was measured. The measurements of thermal durability were performed in the temperature range of a standard recrystallization process (450°C - 650°C). During the experiment, the percentage loss of the foil weight was measured. Additionally, plastic properties were tested as the indicator of usefulness for the elastic substrate application. For higher accuracy of obtained outcomes, as the additional test, the plastic properties of the materials for each temperature were estimated. Complete results of this test are presented in Table 3. Grey color of the table cell marks a permanent deformation or loss of elastic properties.

Weight in temperature:	UPILEX®		KAPTON®	
	12.5µm	25.0µm	12.5µm	25.0µm
480°C	91.82%	95.16%	96.70%	95.30%
500°C	91.36%	94.84%	96.00%	94.60%
550°C	89.55%	92.26%	74.70%	81.12%
600°C	70.00%	78.38%	Burnt	Burnt

Table 3. Temperature durability of examined foils. Dark-grey color indicates the loss of elastic properties or permanent deformation.

Analyzing obtained results, one may state that in the opposite to the manufacturer suggestions, the biggest weight loss in temperatures above 500°C, is observed in polyamide KAPTON®. Additionally, the loss of its elastic parameters occurs very rapidly. Contrary, UPILEX®, which melting point is declared below 400°C proved to be fairly resistant to temperatures until 550°C. In both cases thicker foils reacted slower for the temperature rise, which was expected due to their relatively high thermal resistance. It is worth mentioning that the experiment was conducted in conditions (time, equipment) similar to the manufacturing process. However, identified maximal allowable temperature is relatively lower than standard demanded temperature for ICSVT process. There were reasonable presumptions suggesting the possibility of re-crystallization temperature decreasing, in favor of longer process duration. Thus, examined foils were conditionally positively evaluated. Taking this into account, UPILEX® foil was accepted for further experiments, leading to manufacturing of the CdS/CdTe elastic layers. Considering possible configuration of designed cell, the light transparency characteristic of investigated foil was measured. The light transmission in the conversion range of CdS/CdTe cell both of KAPTON® and UPILEX® foils is presented in Figure 13.

Due to low transmission (below 60%) in the range 400 nm – 700 nm, which would decrease largely the total cell efficiency, for UPILEX®, substrate cell configuration was chosen. Basing on presented results, experimental sample of CdTe absorber, manufactured on 25 µm UPILEX® foil was prepared. Obtained semiconductor layer is based on Cu contact of 2 µm thick, made by PVD in pressure $5 \cdot 10^5$ Torr. The total area of the sample is 30 cm² and elastic properties of all manufactured layers are preserved (Figure 14). After the investigation, the average thickness of 2 µm and good uniformity of manufactured layer was confirmed. This makes proper base for CdS layer manufacturing and completing of the elastic CdS/CdTe construction.

Obtained results confirm the assumption that flexibility of polycrystalline cadmium compound layers may be employed in alternative applications, such as elastic cell structure. Finding the proper material for substrate of these devices is a key to manufacturing of efficient cell, however it demands considering of many technological aspects. Thermal and

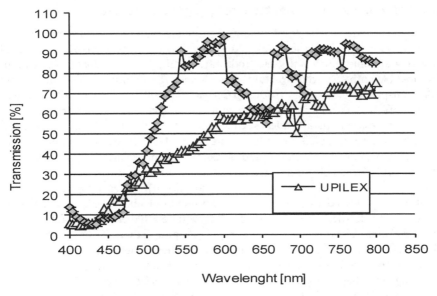

Fig. 13. Optical transparency of KAPTON® and UPILEX® foils in the wavelength range of CdS/CdTe cell effective photoconversion.

Fig. 14. Test structure of elastic CdTe layer based on UPILEX® foil and contacted by 2 μm Cu layer.

mechanical properties of some high-temperature polymer foils give the possibility to construct complete solar cell with some technological modifications (particularly during the re-crystallization process). Another important factor is a proper, flexible and durable contacting system of such cell.

4. Novel carbon nanotube contacts for proposed devices

An essential challenge in the development of flexible photovoltaic structures, excepting the elaboration of an appropriate semiconductor junction and optical properties of active layers, is providing suitable contacts. PV electrodes are required to be reliable, efficient, low cost and compatible with solar cell structure. An extremely frequently used solution is applying flexible transparent conductive oxides (TCO) as PV cell front (generally emitter) layer electrodes. As it was mentioned before emitter contacts are usually realized by using conductive transparent metal oxides, such as: SnO_2, ITO, Zn_2O_4, $CdSnO_4$, In_2O_3, ZnO:Al, as well as CdO, ZnO and $RuSiO_4$. In order to integrate solar cells into PV modules or for more convenient measurements execution, additional metal contacts attached to TCO are applied. The most popular among listed TCO compounds is indium tin oxide (ITO).

ITO is a mixture of tin (IV) oxide: SnO_2 and indium (III) oxide: In_2O_3 so called ITO. This material is characterized by high optical transmission of above 90% in visual range and relatively low electrical resistivity of $10\,\Omega/$square $\div 100\,\Omega/$square for thickness of $150\,nm \div 200\,nm$. Unfortunately, applying ITO and other TCO layers in flexible photovoltaics encountered a significant barrier. Those metal oxides indicate a lack mechanical stress resistance which leads to breaking and crushing of the contact. This disadvantageous characteristic was observed and reported also during the research on flexible diode display electrodes. Furthermore, thin ITO layers are predominantly manufactured by cost-consuming magnetron sputtering method [28], which increases the final cost of new PV cell and module. Moreover, the indium resources are strictly limited and expected to be exhausted within next fifteen years of exploitation.

A novel method of creating flexible transparent contacts for solar cells is to use carbon nanotubes (CNT). Due to the broad range of potential manufacturing techniques and diversified properties of obtained layers, carbon nanotubes are becoming increasingly popular in electronic applications. Especially CNT layers obtained using low-cost technologies such as screen printing or sputtering are potentially useful in flexible electronic devices [30] and smart textiles. This subsection presents the summary of experiments which were conducted up to now and led to adaptation of carbon nanotubes as thin transparent contacts of selected flexible photovoltaic structures.

To create CNT based transparent conductive layer (TCL), preparation of particular composite is necessary. Since there is a requirement of low cost material, multilayer carbon nanotubes, synthesized in catalytic chemical vapor deposition (CCVD), were used in tested compounds. CCVD process has a drawback which causes that not perfectly pure CNT material is obtained. Although, the material contains significant amount of non CNT carbon structures and metal catalyst, either purification or alternative fabrication methods, can increase costs up to a few orders of magnitude. The average dimensions of nanotubes in the material (determined by Scanning Electron Microscopy - SEM) are $10 \div 40\,nm$ in diameter and $0.5 \div 5\,\mu m$ length, however longer structures have also been observed. Figure 15 presents HRSEM image of applied CNTs.

Carbon nanotube composites are printed on given substrates using, low cost screen printing technique. To specify a relationship between the content of CNT in the composition and the value of sheet resistance, electrical properties of printed layers was measured. Table 4 presents achieved results. All samples showed electrical conductivity and were much above the percolation threshold [11].

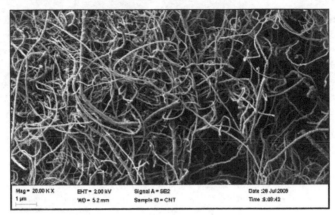

Fig. 15. HRSEM image of applied carbon nanotubes

Paste No	CNT content in the composition [%]	Sheet resistance [Ω/square]
CNT.0.1	0.10	613 k
CNT.0.25	0.25	28 k
CNT.0.5	0.50	3.3 k
CNT.1.0	1.00	870

Table 4. Sheet resistance values for samples with different CNT amount [11].

Transparent conductive layers were prepared using four composites with various CNT content (Table 1). As a substrate borosilicate glass was used. In order to compare CNT and ITO layer parameters, an identical Bo Si glass sample, covered by 160 nm sputtered ITO, was taken. As a first step of carbon nanotubes TCL application in solar cell structure, transmittance of printed layers have been measured (Figure 16).

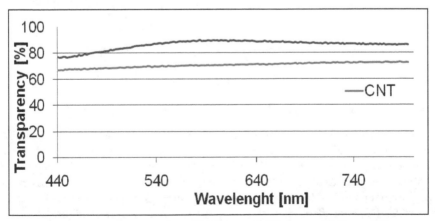

Fig. 16. Transmittance comparison of 0,25%, 1,5 µm CNT layer and 160 nm ITO on borosilicate glass, for standard solar cell absorption spectrum

A very important characteristic for printed CNT layers, is stability of the resistance while applying multiple mechanical stress. To verify this parameter for manufactured CNT layers, additional experiment was undertaken. TCL of 1.5 µm thick was screen printed on polyamide Kapton® and tested by rapid mechanical bending in 80 cycles. The results of resistivity change (Figure 17a) was compared with literature outcomes, obtained for optical ITO layer (Figure 17b).

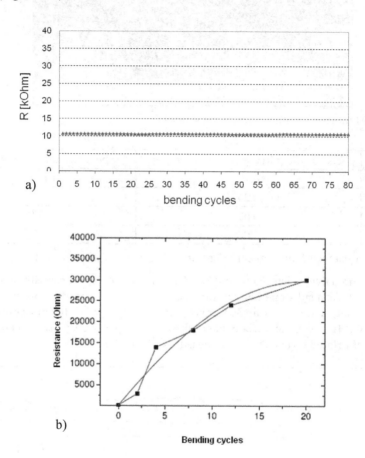

a)

b)

Fig. 17. Resistance changes of: a) CNT and b) ITO layers while bending [31]

After a positive estimation of CNT layers optical and electrical parameters, the possibility of implementation as a solar cell transparent conductive coating was verified. For creating models of screen printed CNT layer, as TCO replacement, in different PV cell structures, SCAPS simulator was used. Simulation models are generated by digital description of physical parameters of each structure layer, including contacts. Solar Cell Capacitance Simulator (SCAPS) is available free of charge for scientific research. Figure 18 shows I-V curves simulations, for CdTe/CdS solar cell structure with ITO and CNT contact layer. Operating parameters of simulated cells are presented in Table 4.

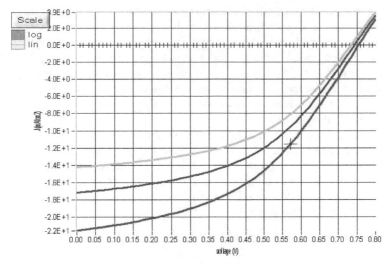

Fig. 18. SCAPS simulations of I-V characteristics of CdTe/CdS solar cell with filters: red-none, blue-ITO, green-CNT.

Filter	Open circuit voltage V_{OC} [V]	Short circuit current J_{SC} [mA/cm²]	Fill Factor FF [%]	Efficiency η [%]
none	0.754	21.602	44.99	7.33
ITO	0.743	17.194	47.00	6.00
CNT	0.733	14.236	48.50	5.06

Table 5. Electrical parameters of CdTe/CdS solar cell

5. Conclusions

Carbon nanotube layers with relatively high optical transmittance were fabricated by inexpensive screen printing technique on glass and on elastic polymer substrates as well. The average difference of 10% in transmittance within standard CdTe cell photoconversion range between 160 nm ITO and 1.5 µm 0.25% CNT layer was observed. Sheet resistance of obtained layers are at relatively high level and should be diminished for efficient photovoltaic applications. To achieve this goal special technology and material compositions (including various CNT content) are tested. The resistance of CNT layers, in opposite to standard ITO, turned out completely independent on bending, which is critical in terms of flexible solar cells construction. According to SCAPS simulations the lowest P_m drop, caused by CNT layer implementation, was observed in case of thin-film cells, which is consistent with postulate of new construction flexibility. Preliminary practical experiments confirmed the presence of photovoltaic effect in solar cell equipped exclusively with CNT emitter electrode.

Presently, due to weaker optical and electrical parameters those layers cannot be a competitive alternative to the existing transparent conductive layers. Nevertheless, they have much better elastic properties and high prospects for improving the optical and electrical parameters, and therefore they can be potential   solar cells layers. Further experiments are planned for development of manufactured structure (including incorporation of main metal contacts) and manufacturing of thin-film cells with carbon nanotube emitter contacts. However, CNT composites obtain higher optical permeability at a lower carbon nanotubes content, which in turn, increases the resistivity of these materials. Thus, the simultaneous increasing of the permeability and reducing the resistivity is a difficult issue.

Flexible solar cells, based on thin film heterostructure are expected to be a natural development of currently produced devices. For elaboration of fully functional photovoltaic structure, ready for industrial production, many technological problems must be solved. Presented work is a small part of impact put in this process. It is highly probable that some of presented concepts will soon find the implementation in the commercially available elastic cells, based on II-VI compounds.

6. References

[1] K. Zweibel "Thin Films: Past, Present, Future" Progress in Photovoltaics, Special Issue on Thin Films, NREL 1995.

[2] O. Mah "Fundamentals of photovoltaic materials" National Solar Power Research Institute 1998 pp 1-10.

[3] A. Hepp et al. "Ultra-Lightweight Hybrid Thin-Film Solar cells: A survey of Enabling Technologies for Space Power Applications" Proc. 5th International Energy Conversion Engineering Conference and Exhibit (IECEC) St Louis 2007 p 4721.

[4] V. Bemudez, A. Moreau, N. Laurent, L. Jastrzebski "Roadmap of characterization techniques for the development of thin film photovoltaic technology" Proc. Photovoltaic Technical Conference – Thin Film 2010, Aix-en-Provence, France 2010.

[5] D. Bonnet, H. Rabenhorst "New results on the development of a thin film p-CdTe/n-CdS heterojunction solar cell" Proc. 9th IEEE Photovoltaic Specialist Conference, New York 1972 pp 129-131.

[6] X. Wu, J. Keane, R Dhere, C. DeHart, A. Duda, T. Gessert, S. Asher, D. Levi, P. Sheldon „16.5% efficiency CdS/CdTe polycrystalline thin film solar cell" Proc. 17th European Photovoltaic Solar Energy Conference, Munich 2002 pp 995-1000.

[7] ZSW Press Release 05/2010, Stuttgart, Germany 2010.

[8] A. Tiwari "Flexible solar cells for cost effective electricity. High efficiency flexible solar cells based on CIGS and CdTe" Proc. Photovoltaic Technical Conference – Thin Film 2010, Aix-en-Provence, France 2010.

[9] V. Fthenakis, EMRS-2006 Spring meeting

[10] J. Perrenoud, B. Schaffner, L. Kranz, S. Buecheler, A. Tiwari „Flexible CdTe thin film solar modules" Proc. Photovoltaic Technical Conference – Thin Film 2010, Aix-en-Provence, France 2010.

[11] M. Sibiński, M. Jakubowska, K. Znajdek, M. Słoma, B. Guzowski „Carbon nanotube transparent conductive layers for solar cells applications", Proc. 10th Electron

Technology Conf. ELTE 2010 and 34th International Microelectronics and Packaging IMAPS-CPMT Conf., 22-25.09, 2010, 81-82.

[12] H. S. Ullal, K. Zweibel, B. G. Roedem "Polycrystalline thin-film photovoltaic technologies: from the laboratory to commercialization" NREL 0-7803-5772-8/00 IEEE 2000.

[13] H.S. Ullal, B. Roedern "Thin Film CIGS and CdTe Photovoltaic Technologies: Commercialization, Critical Issues, and Applications" Proc. 22nd European Photovoltaic Solar Energy Conference (PVSEC) and Exhibition, Milan, Italy 2007

[14] P. Mints „Principal Analyst Navigant Consulting" PV Services Program 2010

[15] T. Markvart, L. Castaner "Solar Cells: Materials, Manufacture and Operation" Elsevier Amsterdam 2006.

[16] T. Nisho Thin film CdS/CdTe solar cell with 15,05% efficiency 25 th Photovoltaic Specialists Conference 1996 ss. 953 -956

[17] L. Kazmierski, W. Berry, C. Allen „Role of defects in determining the electrical properties of CdS thin films". J. Appl Phys Vol43 No8 1972 pp 3515-3527

[18] R. Bube „Photovoltaic materials" Imperial college press Londyn 1998 pp 135-136.

[19] S. Bernardi "MOCVD of CdTe on foreign substrates". Materials Science Forum Vol 203 1996 ss115-122.

[20] C. Ferekides, D Marinski, V. Viswanathan i in." High efficiency CSS solar cells". Thin Solid Films 2000 ss 520-526

[21] Mendoza-Pérez, R.b , Aguilar-Hernández, J.R.a , Sastré-Hernández, J.a , Tufiño-Velázquez, M.a, Vigil-Galán, O.a , Contreras-Puente, G.S.a , Morales Acevedo, A.c , Escamilla-Esquivel, A.a , Ortega-Nájera, B.a , Mathew, X.d , Jean-Marc-Zisae "Photovoltaic modules processing of CdS/CdTe by CSVT in 40 cm2" 2009 34th IEEE Photovoltaic Specialists Conference, PVSC 2009; Philadelphia

[22] M. Sibiński, M. Burgelman „Development of the thin-film solar cells technology". Microtherm '2000 Łódź-Zakopane 2000 ss. 53-60.

[23] B. Depuydt, I Clemminck, M. Burgelman, M. Casteleyn, . "Solar Cells with screen-printed and sintered CdTe layers on CdS/TCO substrates". Proc. of the 12th EPSEC Stephens & Associates. 1994. ss. 1554-1556

[24] M. Sibiński, Z. Lisik "Polycrystalline CdTe solar cells on elastic substrates", Bulletin of the Polish Academy of Sciences, Technical Sciences Vol. 55, No. 3, 2007, 2007, str. 287-292

[25] M. Sibiński "Thin film CdTe Solar Cells in Building Integrated Photovoltaics", 1st SWH International Conference, 13-15 (2003).

[26] I. Lauremann, I. Luck, K. Wojczykowski „CuInS2 based thin film solar cells on roof tile substrates" 17th EPSEC 1256-1259 (2001)

[27] D. Batzner, A. Romeo, D. Rudman, M. Kalin, H. Zogg, A. Tiwari . " CdTe/CdS and CIGS thin Film Solar Cells." 1st SWH International Conference 56-60 (2003),

[28] A. Hepp et al. "Ultra-Lightweight Hybrid Thin-Film Solar cells: A survey of Enabling Technologies for Space Power Applications" 5th International Energy Conversion Engineering Conference and Exhibit (IECEC) St Louis 2007 p 4721

[29] H. Sirringhaus, T. Kawase, R. H. Friend, T. Shimoda, M. Inbasekaran, W. Wu, E. P. Woo, "High-Resolution Inkjet Printing of All-Polymer Transistor Circuits", Science 15, vol. 290, no. 5499, 2000, pp. 2123 – 2126.

[30] Y. Seunghyup, Y. Changhun, H. Seung-Chan and C. Hyunsoo *"Flexible/ ITO-free organic optoelectronic devices based on versatile multilayer electrodes"* – Raport Integrated Organic Electronics Lab (IOEL)Dept. of Electrical EngineeringKorea Advanced Institute of Science and Technology (KAIST), Daejeon, Korea 2009.

Permissions

The contributors of this book come from diverse backgrounds, making this book a truly international effort. This book will bring forth new frontiers with its revolutionizing research information and detailed analysis of the nascent developments around the world.

We would like to thank Professor, Doctor of Sciences, Leonid A. Kosyachenko, for lending his expertise to make the book truly unique. He has played a crucial role in the development of this book. Without his invaluable contribution this book wouldn't have been possible. He has made vital efforts to compile up to date information on the varied aspects of this subject to make this book a valuable addition to the collection of many professionals and students.

This book was conceptualized with the vision of imparting up-to-date information and advanced data in this field. To ensure the same, a matchless editorial board was set up. Every individual on the board went through rigorous rounds of assessment to prove their worth. After which they invested a large part of their time researching and compiling the most relevant data for our readers. Conferences and sessions were held from time to time between the editorial board and the contributing authors to present the data in the most comprehensible form. The editorial team has worked tirelessly to provide valuable and valid information to help people across the globe.

Every chapter published in this book has been scrutinized by our experts. Their significance has been extensively debated. The topics covered herein carry significant findings which will fuel the growth of the discipline. They may even be implemented as practical applications or may be referred to as a beginning point for another development. Chapters in this book were first published by InTech; hereby published with permission under the Creative Commons Attribution License or equivalent.

The editorial board has been involved in producing this book since its inception. They have spent rigorous hours researching and exploring the diverse topics which have resulted in the successful publishing of this book. They have passed on their knowledge of decades through this book. To expedite this challenging task, the publisher supported the team at every step. A small team of assistant editors was also appointed to further simplify the editing procedure and attain best results for the readers.

Our editorial team has been hand-picked from every corner of the world. Their multi-ethnicity adds dynamic inputs to the discussions which result in innovative outcomes. These outcomes are then further discussed with the researchers and contributors who give their valuable feedback and opinion regarding the same. The feedback is then collaborated with the researches and they are edited in a comprehensive manner to aid the understanding of the subject.

Apart from the editorial board, the designing team has also invested a significant amount of their time in understanding the subject and creating the most relevant covers. They scrutinized every image to scout for the most suitable representation of the subject and create an appropriate cover for the book.

The publishing team has been involved in this book since its early stages. They were actively engaged in every process, be it collecting the data, connecting with the contributors or procuring relevant information. The team has been an ardent support to the editorial, designing and production team. Their endless efforts to recruit the best for this project, has resulted in the accomplishment of this book. They are a veteran in the field of academics and their pool of knowledge is as vast as their experience in printing. Their expertise and guidance has proved useful at every step. Their uncompromising quality standards have made this book an exceptional effort. Their encouragement from time to time has been an inspiration for everyone.

The publisher and the editorial board hope that this book will prove to be a valuable piece of knowledge for researchers, students, practitioners and scholars across the globe.

List of Contributors

Leonid A. Kosyachenko
Chernivtsi National University, Ukraine

Verka Georgieva and Marina Georgieva
Faculty of Electrical Engineering and Information Technology, The "St. Cyril & Methodius" University, Skopje, R. of Macedonia

Atanas Tanusevski
Faculty of Electrical Engineering and Information Technology, Institute of Physics, Faculty of Natural Sciences and Mathematics, The "St. Cyril & Methodius" University, Skopje, R. of Macedonia

Shuo-Jen Lee and Wen-Cheng Ke
Yuan Ze University, Taiwan, R.O.C.

Ruwan Palitha Wijesundera
Department of Physics, University of Kelaniya, Kelaniya, Sri Lanka

L. Fu
College of Materials Science, Northwestern Polytechnical University, Xian, State Key Laboratory of Solidification Processing, P. R. China

Fritz Falk and Gudrun Andrä
Institute of Photonic Technology, Germany

Z.Q. Ma
SHU-Solar E PV Laboratory, Department of Physics, Shanghai University, Shanghai, P. R. China

B. He
Department of Applied Physics, Donghua University, Shanghai, P. R. China

Antonio De Luca
VEGA Space GmbH, Germany

Jongho Yoon
Hanbat National University, Republic of Korea

Maciej Sibiński and Katarzyna Znajdek
Technical University of Łódź, Department of Semiconductor and Optoelectronic Devices, Poland

Printed in the USA
CPSIA information can be obtained
at www.ICGtesting.com
JSHW011424221024
72173JS00004B/666